William A. Cook

The Horse

Its keep and management

William A. Cook

The Horse
Its keep and management

ISBN/EAN: 9783744648318

Printed in Europe, USA, Canada, Australia, Japan

Cover: Foto ©berggeist007 / pixelio.de

More available books at **www.hansebooks.com**

THE HORSE:

ITS KEEP & MANAGEMENT

BY

WILLIAM COOK,

/I

Author of "The Practical Poultry Breeder and Feeder: or, How to make
Poultry pay ;" " The Book on Ducks, and how to make them pay ;"
Editor and Proprietor of " The Poultry Journal ;" Conductor of
Poultry Department " Farm, Field and Fireside;" Weekly Contributor
to " Poultry," &c., &c.

PRICE 2s. 6d.

PUBLISHED BY THE AUTHOR AT

ORPINGTON HOUSE, ST. MARY CRAY, KENT,

AND AT

QUEEN'S HEAD YARD, 105, BOROUGH, LONDON, S.E.

ENTERED AT STATIONERS' HALL.

1891.

OPINIONS OF THE PRESS.

Even the most inexperienced in matters equine cannot fail to become pretty well acquainted with the practical management of the stable by perusing Mr. W. Cook's work, *The Horse: Its Keep and Management.* The work, published by the author, is written with the humane object of lessening the sufferings of horses, caused as often by the ignorance of those who have charge of them as by cruelty. Mr. Cook treats his subject in a comprehensive manner, and writes in a clear and concise style. —*Lloyd's*, August 25th, 1891.

The author, William Cook, of *The Horse: Its Keep and Management*, which he publishes himself from Orpington House, St. Mary Cray, Kent, is a horse-breeder, and, what is more, evidently a lover of the noble animal. His object in writing this little book is to make the lives of these useful animals happier by making those who have the control of them understand them better. He treats of their feeding, watering, manipulation in the stable, their ailments, breeding, breaking-in, driving, &c. ; and he does it in such a simple unaffected way that he who runs may read. If the kind of information here given were disseminated amongst drivers and stablemen and all those who have the handling of horses, it would have a more meliorating effect upon the lot of the equine race than any number of police-court prosecutions. Take our 'bus horses alone. How often when one of them has stumbled may the driver not be observed to give it a lash and begin to jerk ferociously at the reins? The time to give it a touch with the whip and to stiffen the reins—not jerk them—is when there is a likelihood of its stumbling. To cut it after it has stumbled is, as Mr. Cook remarks, to make it more frightened next time it stumbles, as it expects the lash, and "in its excitement and nervousness its knees are on the road before it knows where it is." Another point touched on by the author, which we consider well worth emphasising, is the treatment of a horse that has fallen. To get him up the driver frequently proceeds to kick him on the ribs or to apply the whip. "Now," says Mr. Cook, "this is most inhuman; the horse would not fall down if he could help it. That is not all, however ; when he is down he is very frightened and nervous, but when it comes to lashing him with the whip, and kicking him, in less than three minutes the poor thing is all of a lather with fright. Any man who does this kind of thing ought to have six months' hard labour ; yet it is done every day." No ; the way to act is to be calm, collected, and kind ; get him quietly disentangled from the harness and vehicle, some one holding his head down the while, and then he will rise. No amount of teaching, however, will, we fear, make good drivers out of badly-dispositioned men ; for it is to a certain extent true of the handlers of horses as it is of poets—"they are born, not made."—*Daily Chronicle*, August 26th, 1891.

The Horse: Its Keep and Management (By William Cook).
—This volume has the very great merit (and it must be added
the corresponding defect) that it represents only the opinion of
one man.   Still he is an intelligent observer of the horse in its
work and in its stable.   There are many much more comprehensive volumes on the horse, and books in which the whole
subject is more exhaustively treated ; but Mr. Cook tells what
he believes himself to have learned by his own experience,
and his experience really has been considerable.   We do not
say a man's experience—let his life be ever so long and his field
of observation be ever so wide—can tell him all that ought to
be known in order to discuss thoroughly any subject.   But no
amount of reading on such a question as this of how to keep a
horse, can compensate for the want of experience.   The chapters
which tell of feeding and stable-management are full of useful
hints to the owner of one horse.   The other chapters are more
ambitious and less successful.   But even the first chapters alone
are very good value for half-a-crown to those who for the first
time are about to start a horse and trap of their own.   Mr.
Cook never leaves one in doubt as to what his meaning is.
His advice is lucid enough.—*Live Stock Journal*, August 28th,
1891.

The Horse: Its Keep and Management, by Mr. William
Cook, is a sensible and practical little book, the result of considerable observation and experience.   It may be taken as a
guide by all those whose acquaintance with the subject is
limited, while many claiming special knowledge will find
numerous hints and suggestions, the adoption of which would
prove profitable to them and beneficial to their horses.   The
horse is naturally intelligent and docile, though playful and
spirited, and responds readily to kindness, by the agency of
which he can generally be made to obey his master's wishes
without having recourse to punishments which injure his
temper and his nerves.   The burden of the book is to be gentle
and patient, to treat a young horse much as you would a
child, and when you do punish to be influenced by reason, and
not by temper.   Mr. Cook, of Orpington, is himself the publisher
of the manual.—*Morning Post*, September 2nd, 1891.

The Horse: Its Keep and Management.—The author of this
work (Mr. Wm. Cook) needs no introduction to Poultry keepers,
but perhaps some may fancy that it is a big leap from Poultry
keeping to the horse and its management ; nevertheless we can
assure readers that such is not the case here, as Mr. Cook has
been amongst horses all his life, and is as thoroughly acquainted
with them as with Poultry.   The work in question does not
attempt to deal in any scientific manner with the question.   It
simply deals with the practical management in such a way as
will enable the most inexperienced to find in it plain and
practical hints. The subjects, amongst others, dealt with include

feeding, watering, bedding, cleaning, clipping, singeing, shoeing, breeding, breaking, driving, diseases, &c., &c. It is written in plain simple language, that he who runs may read, and thousands who may have been stranded with scientific works cannot fail to find in this work much sound information. It is written by a shrewd observer, and will be found of very great service to those to whom it appeals.—*Poultry.*

THE A.B.C. OF HORSE MANAGEMENT.—Mr. W. Cook, of Orpington House, St. Mary Cray, who has already achieved some degree of fame as an exponent of the art of poultry-keeping so as to pay, has turned his attention to the treatment of horses, and has issued a simple handbook to horse manage-ment which should be of more use to the inexperienced in such matters than the highly scientific compilations on the same subject, used by more advanced folk. The author confesses in his preface that his object in view is not so much the examina-tion of scientific theories on the subject as simply to deal with the practical management of the horse in such a way as to enable the most inexperienced to find in its pages the plain and simple facts of matters which press for attention, and sufferings which call for speedy relief. During his travels in various parts of the country he has been "frequently appalled" at the bad management to which horses are subjected. He is inclined to think that this is not the result of a desire to be cruel on the part of those who keep the animals, but the outcome often-times of ignorance as to their real wants, and the treatment that is necessary and effectual to the relief of their sufferings. Having therefore a claim to be heard as having been among horses from a very early age, he has written the book, and has sent it forth to the public with a sincere desire that many will find it helpful, and that the horses of this country will, as a result of its appearance, be treated better and more intelligently. The book itself contains chapters on feeding, watering, bedding-down, and on breeding, breaking-in, horses' diseases, and subjects of more general information. It is nicely turned out by Messrs. E. Clarke and Sons, of St. Mary Cray, and although it contains many statements open to controversy, it is likely to prove useful.—*District Times*, August 8th, 1891.

The Horse: Its Keep and Management (By William Cook):— In homely style Mr. Cook has indited a book which will not be without its value to those who keep nag horses for purely utilitarian purposes. In justice, however, to a class of men who are much maligned, we feel bound to protest against such sweeping passages as this: "It is a known fact horse dealers are looked upon as a race of men who will tell any untruth as long as they are able to strike a good bargain." Again, "I . . . find in many cases a horse dealer will often sell a gentleman a horse, telling him he is a first-class animal and will just suit him, when the man knows all the time it will be of no use to the purchaser." This is tantamount to a charge of deliberate fraud. Some dealers there are whom no one would trust, just as there are tailors, bootmakers, grocers, and others of whom every sensible man would beware; but given men of reputation, and they are neither more nor less honest than other tradesmen; while many act in strictly honourable fashion towards their customers.

Like a good many more persons, Mr. Cook has managed to mixed up intelligence with instinct, and calls a horse intelligent because, having once stopped at a certain place, he looked out for being pulled up there when he next passed, though a twelve-month had elapsed between the two journeys. This, we venture to suggest, was not intelligence—it was memory, the horse's strong point, upon which every skilful trainer and breaker plays.

In the remarks on feeding and watering horses there is a good deal of common sense; but Mr. Cook would have done better to have given more definite rules for the guidance of the inexperienced. The little-and-often principle is good up to a certain point: but four feeds a day we consider ample. We are quite at one with Mr. Cook in his recommendation of frequent watering; and we cordially agree with him in his condemnation of soft food for horses. The author is not a violent advocate for the washing of horses' legs; but we would go further, and say that under no circumstances should the legs of a hunter be washed; friction, and not water, is needed to cleanse the hair and skin, and there is no greater promoter of mud fever than washing the legs. Mr. Cook's work may be recommended to those who are about to set up a useful pony or cob, as the inexperienced horse owner will read several things that are worth remembering; and, as the book is couched in very homely language, it will be the better understood. Those who keep hunters will not find much dealing with that particular class of horse; nor is anything said about the superior class of carriage horses. As a manual for general principles, the book may be read with profit.—From *The Field—Horse Management.*

The Horse: Its Keep and Management (Mr. William Cook).— Author of "The Practical Poultry Breeder and Feeder," &c. Published by the Author, Orpington House, St. Mary Cray, Kent. Price 2s. 6d. It is evident that Mr. Cook has written this book *con amore*, he having been amongst horses from a very early age. He tells in a very homely way how the animal should be bred, fed, broken, and treated, and keeps entirely away from the scientific side of the subject. What he recommends is the outcome of experience; and if the volume is less presumptuous than many others we come across, it can be understood by those who read it, and numerous useful hints may be obtained from it.— From *Farm, Field & Fireside.*

The Horse: its Keep and Management (By William Cook):—Anyone who writes a Book with the avowed intention of teaching the British public how to treat their horses **better than is, alas! the present rule deserves the sympathy** of all lovers of the noble quadruped, and their support in the exact proportion of his ability to impress wholesome instruction upon the more ignorant of the community, from whom undeniably horses suffer many wrongs—some intentionally inflicted, but many the outcome of mere thoughtlessness.

The author of the above named volume has evidently, as he himself states, spent much of his time in the study of his subject, both from practical and theoretical points of view. And we can heartily commend many of his tenets. It is, no doubt, as he says, "not always well even to scold horses when they do wrong ; horses are much more intelligent than most people take them to be ; " and an angry word upon a slight occasion may, upon the next, render the occasion one of real danger, from the sudden memory of the previous rating causing the horse to lose his nerve, and do something foolish to avoid the recurrence of the well-remembered reproof. A shying horse, for instance, will add the reproof or the cut of the whip which he received upon the last offence to his, perhaps, real dread of the object at which he shied, and probably, as the author says, "do something desperate to escape this double source of uneasiness," or a horse who has been struck for stumbling will naturally start off at a gallop after the next mistake of the kind, to avoid the expected punishment ; and a frightened gallop means danger. At the same time Mr. Cook is not wholly averse to punishments when salutary, as, for example, when he most rightly advises prompt reproof, and a cut with the whip "when a horse throws back its ears and lifts its foot up in the stable to kick," instead of the attendant running out of the way, as too often happens with half-taught grooms, the surest way to convert the threat or play into a real kick upon the next occasion. Any temper in adminis- tering such punishments should, however, be avoided, and one blow is enough.

The author's remarks upon breeding, etc., are much to the point, notably two practical hints, the first being a warning against backing mares in foal when in harness, as a frequent cause of foal slipping ; and the second as to the advantage and size derived by colts from a liberal diet of cow's milk.—*Land and Water*, August 29th, 1891.

The Horse: Its Keep and Management (By William Cook). *Farm and Home*, after many complaints says : The recom- mendations as to management are decidedly valuable, and the truth is told as to the general mismanagement as regards feeding, the use of cooked food, watering, bedding, and driving, in terms that should make the average Englishman, with his inborn conceit about all that appertains to matters equine, positively "squirm." There is a ring of genuine originality about the book : it is obviously founded on experience which has led the author into drawing correct conclusions that accord with scientific teaching as well as with practical kuowledge, and there is absolutely no "crib" or borrowing of other people's ideas.— From *Farm and Home*, October, 10th, 1891.

The Horse: Its Keep and Management, by William Cook, (Orpington, St. Mary Cray). Price 2s. 6d. It is with very great pleasure that we have noticed the large number of publications which have recently made their appearance in connection with the horse and its management. It is a healthy sign of the times when so many qualified writers contribute their valuable opinions upon so interesting a subject. Amongst the large number of books that have within recent months made their appearance, none, we venture to say, will meet with a heartier welcome than that now under notice. Mr. William Cook has already established a reputation as a writer in connection with books on poultry, and we know of no one who is better qualified to write upon the subject of the horse than this gentleman. Mr. Cook does not seek to present us with any scientific theory ; he is quite content to put forward a simple and easily-understood treatise upon the practical management of the horse, which "those who run may read." The author has travelled far and wide throughout our own country, and he has noticed what a great many others who have kept their eyes open must have equally observed— namely, the abominable and cruel treatment to which some horses are subjected. No doubt he has often blushed, as have we, at the simple barbarity displayed by many persons who keep animals ; and his sole object is to improve the conditions under which the horse has to live. From his very earliest age he has been much amongst horses, and has naturally made himself fully acquainted with their habits, requirements, and sufferings. We can cordially recommend this book to our readers, and we have been so favourably impressed with it that we shall hope to see other publications of a similarly useful nature from Mr. William Cook's able pen before long. — From *The Road*, October 1st, 1891.

The Horse: Its Keep and Management (By W. Cook).— The author of this handy manual is the well-known originator of the Orpington fowl, and as such, requires no introduction to poultry keepers, but from poultry-keeping to horse-management may seem to some people a long jump indeed ; though, in this instance, we can assure our readers that it is not so. Mr. Cook has been amongst horses all his life, and is as thoroughly acquainted with them as with poultry. The work in question does not attempt to deal in any scientific manner with the question ; it simply deals with the practical management of the horse in such a way as will enable the most inexperienced to find in it plain and practical hints. The subjects include feeding, watering, bedding, clipping, singeing, shoeing, breeding, breaking, driving, diseases, &c., &c. It is written by a shrewd observer, and should be in the hands of all lovers and keepers of the most useful of all animals.—*Finchley and Hendon Times*, August 14th, 1891.

CONTENTS.

INDEX.

ILLUSTRATIONS.

PREFACE.

PERHAPS in presenting my book on "The Horse" to the public I may be allowed to offer a few remarks concerning the object I have in view; which is not looking so much to scientific theories, as simply dealing with the practical management of the horse in such a way as to enable the most inexperienced to find in its pages the plain and simple facts of matters which press for attention, and sufferings which call for speedy relief.

During my travels in various parts of the country I am frequently appalled at the bad management to which some horses are subjected. This I am inclined to think is not the result of a desire to be cruel on the part of those who keep the animals, but is the outcome oftentimes of ignorance as to their real wants and the treatment that is necessary and effectual to the relief of their sufferings.

Having been among horses from a very early age, and being interested in them, I have naturally learned all I could of their habits, needs, and sufferings; and because I like so much to see horses treated so that they are happy and well, I have written this book, and send it forth to the public with a sincere desire that many will find it helpful, and that the horses of this country will, as a result of its appearance, be treated better and more intelligently; and also that in some instances means will be employed in the care and management of horses, whose lives have perhaps been a burden to them through weakness and low condition, resulting from inattention and bad treatment, which will bring them brighter days and a happier existence.

WILLIAM COOK.

ORPINGTON HOUSE,

ST. MARY CRAY, KENT.

General Remarks and Information.

The writer's object—How horses are spoiled—The horse dealer—
beware!—The folly of using harsh measures, and injurious after-
effects resulting therefrom—Shying and making shy.

MANY horses in this country are not only abused
by those who attend to them, but in many cases
by the owners themselves. It is the mismanagement of
these noble animals which has induced me to give a
little spare time to writing this little book. I am neither
a veterinary surgeon nor a horse dealer—but a horse
breeder, and my intention is just to give a general out-
line of the management of the horse, as I am a lover
of the animal. To experienced persons this work may not
be of any very great value, but even those who have
had long experience and are really very fond of their
horses will no doubt find something useful. The book
is written in so simple a form that even a boy can
understand it. I usually find it is far better to put
things in such a way as though no one knew anything
about the subject but the writer. This enables practical
persons to see they are right, and those who are not
can just have a bird's-eye view of the most simple way
to manage a horse or horses.

B

I have treated upon feeding horses so that inexperienced people will know how to feed their animals when they are either hard at work or standing still in the stable. In the following chapters will be seen what to feed the horse upon, and how the food should be given. I have also treated upon the breaking-in of young horses—how they can be tamed down and gently brought into work with but very little trouble, as well as how a restless horse can be taught to stand quiet almost directly it is broken-in. I have also endeavoured to point out how a horse can be driven on a long journey without injuring the animal in any way.

There is a chapter on breeding—stating how many valuable mares are practically lost when they get to a certain age, or, in other words, how many good mares are sold for a trifle, which, with very little trouble and forethought, might be sent to some farmer in the country and bred from. This can easily be done. In some instances a colt can be reared for from £35 to £40 up to when it is four years old, and if from an extraordinarily good mare it sometimes turns out worth from £70 to £100.

It is needless for me to say how tricky horse-dealing is in this country, as it is a known fact horse-dealers are looked upon as a race of men who will tell any untruth so long as they are able to strike a good bargain. But even that is not the worst of it. They often sell a person a horse which will either kick or jib—perhaps both—and so endanger the lives of many inexperienced persons. Horse buying is generally an awkward matter even to those who understand the animals, but worse for those who are inexperienced as to the way they can be got up for sale.

I have been very much struck during the last few years with the number of accidents that happen where people have met with their death, and others have been injured. In many cases it has been simply through having a fresh horse and not understanding it. In other words, the animal was not what he was represented to be. I do not mean to say there are no honest horse dealers, but I do say this—they are few and far between in comparison to the number of dishonest ones. I have had a great deal to do with horses during my life, and find in many cases a horse-dealer will often sell a gentleman a horse, telling him he is a first-class animal and will just suit him, when the man knows all the time it will be of no use to the purchaser. The dealer will often go so far as to send another man to buy the horse back, of course, for less money, and will send the gentleman a worse one in its place. No matter what neighbourhood people go into, if they are thinking of buying a horse they scarcely dare tell anyone, or they will be tormented by a lot of horse-dealers. The latter are not wise in their day and generation, if they were they would treat people honestly.

There is a splendid opening for honest horse-dealers in our country at the present day. Suppose, for instance, a gentleman buys a horse of a dealer, and the animal turns out to be well worth the money, the owner's friends at once ask where it came from, and in many cases a good order is the result : by this means the dealer is recommended from one to the other. This is how things should be, then one need not fear to buy a horse, even though he may not know much about them himself. As I travel much in almost every county in England during the twelve months, I see a great deal of changing hands with horses, but I prefer breeding my own, unless I can get them direct from the breeder or farmer. I am determined never to have anything to do with horse-dealers, if I do I shall want the animal a month on trial. When a horse is being sold I am sure it would be much better if the person selling told the intending purchaser the faults of the animal. When I have one for sale, I always do this, but it is very seldom I have one to spare.

Horses are often abused because those who drive them are accustomed to stop outside public-houses. It often grieves me when I see poor horses on a very wet day standing shivering outside these places. If horses could tell tales what strange stories many of them could relate. Though they are dumb animals they are wonderfully intelligent. Many young horses are spoiled in breaking-in, while others are ruined after they are broken-in. If a horse is broken-in properly it seems to know almost every word that is said to it.

Horse-breeding is largely on the increase in England, and from the last six to ten years horses have been wonderfully improved. This is owing, to a certain extent, to the number of shows which have been held in different counties in England. Hunters, nag horses, and heavy cart horses are all of far better quality than they used to be.

As in my younger days I had a good deal to do with breaking-in horses, I have a good idea as to how young horses should be handled. The few observations I have made during my life have been very helpful to me, as I not only learned the best way to break-in colts, but the way to manage them after they were broken-in. I have not recommended any particular bit being used for horses, as I seldom use anything but a "snaffle" or a Liverpool bit. It is very necessary when a person is buying a horse he should know what kind of bit the animal is accustomed to, as, if a horse is nervous and gets to a strange place, has a fresh driver, and a totally different bit to what he has been accustomed to, it frightens him very much, and very often the owner also. Those who rear young horses should handle them a little more when they are colts, if this were done they would be much less trouble afterwards. Many people use both a bit and a curb for young horses. This is done to make them shape the neck well, which, of course, sets a horse off, and in some people's estimation puts a "£10 note" on it. Then again, there are others who rein them up very tightly, which often really amounts to cruelty. I frequently see a horse's mouth bleed through being reined up too tightly. This is a mistake which many people make. I will give my reasons why I do not believe in doing this. When a person is driving a horse which is treated in this way the driver has not so much power over the animal, as it has a certain weight on the bearing rein. Not only that, but it tends to numb the horse's mouth. When the animal is going along and tosses his head to and fro, the driver's hands give to the bit, but not so with the bearing-rein, that is a certain length and does not give.

I have been noticing particularly the number of horses which have fallen down—not old crocks, but animals which have done good service,—and I find by far the greater majority of these are reined up tightly.

With a young horse, however, it is quite different. I know some people object to using a bearing-rein, but my idea is a young horse ought to have one, especially an animal which is going to be used for some purposes. If a bearing-rein is not used for a horse that stands about, it gets accustomed to putting its head down and biting the grass off the side of the road, that is if the animal happens to be in a country place, and if in a town it wants to smell the road. When putting the head down it will often get the shaft under the collar, which in many cases results in an accident. Young horses will get up to many little pranks of this kind unless they are very quiet indeed. Now the bearing-rein teaches the horse to hold up its head and will often prevent accidents, but in no case should the rein be too tight. It is a different matter when it is a gentleman's horse, and is either driven by the owner or groom, such an animal should never be left alone, and in a case of this kind I prefer not using a bearing-rein, unless it is just for a few weeks when the horse is first driven, then it should be very slack indeed. I find in almost every case the animal has a better mouth when not held up with the bearing-rein. Another thing, a horse yields so much better to the touch of the bit when broken-in without using the former.

When a horse stumbles, harsh measures should never be resorted to. This is a mistake which is often made. Nineteen drivers out of every twenty, when a horse slips, will give the poor thing a good lash with the whip. When it stumbles the next time the animal is doubly frightened, as it expects a sharp cut with the whip, and in its excitement and nervousness its knees are on the road before it knows where it is. If a horse is likely to stumble, it should be just touched up with the whip, and kept well in hand, but not just when it makes the slip.

It is not always well even to scold horses when they do wrong. I will give my reasons why. When a horse is standing about and happens to make a mistake, and commences to bite the hedge, gets hold of a little bit of grass, or puts its head over a fence,—which a young mischievous horse will do sometimes,—the owner or attendant usually gives the animal's head a good snatch with the rein, and commences scolding at the same time, so that the horse gets to know its driver's sharp voice. What is the result? Should the rein get caught in the fence, or hung up in any way, or when the animal is in close quarters, and knows it has done wrong, it will run back and sometimes break the reins, or perhaps get its bridle off: the end is—a runaway and smash-up. If it does not do this it often backs into something, and the smash and rattle of the cart so startle the animal that if someone is not near at hand a great deal of mischief is done. I have seen people not only snatch the horse's head but give it a good cut with the whip when it does wrong. This is a most dangerous process. When I have left my horse a little while, and the rein has got caught somewhere, I find a kind word will usually keep the animal as quiet as possible. I have been in dangerous places before now, and perhaps a trace or some part of the harness has given way, but when I have just called out, "Whoa, my boy!" "Whoa, my beauty!" it has stopped my horse or horses at once, and no harm has come of it. Horses should be so trained that in the event of any part of the harness breaking or other accident occurring they will stand still at once.

Those which I have broken-in (though that is not very many), would come to me if I called them with less than three months' training. I have tried this by taking them out into the middle of a field with a saddle and bridle on and calling to them

Horses are much more intelligent than most people take them to be. Their instinct is wonderful. I remember driving a horse on a strange road some years ago, and I stopped at a cottage to ask the way. To the very day twelve-months, after I went the same road again, and some distance before the animal got to the cottage, he pricked up his ears, and looked up, as much as to say—this is where I stopped when I was this way before. As soon as we got to the cottage, my horse stopped within a foot of the same place as he did the year before. Now I have had scores of such instances as this come under my notice with horses under my charge. I have been accustomed to take them for very long journeys—300 miles at a stretch—and in many cases where I have been over the same road the second time the horse has pulled up again at the same places as he did before.

Sometimes it has been three or four years before I have gone the same road again, and even then, in some instances, the horse would take the same turnings he did when he went the same way before. This seems very wonderful, and seems to denote an extraordinary amount of intelligence.

Now take a horse, for instance, at work on the farm. If he has been accustomed to being loosed from the cart or plough at a certain time in the day he will know almost to the minute when he is going to be set free, just the same as he knows when feeding time comes round. Try another plan—go into a field where a pet

horse is lying out with a number of others, let the
owner of the pet animal call it, and if its master is with
a dozen other people standing in a row, the animal's
instinct tells him immediately to whom he belongs? When a
horse is trained to do certain things it knows what is
said to it almost the same as human beings. I believe it
is a pleasure for a horse to do its work when the
owner or attendant is kind to it. I always like to see
horses put their ears forward rather than throw them
back, because when they raise them it shows they
are pleased, but when they lay them back it is the
contrary, it shows they are displeased. This can easily
be seen by giving a horse a piece of bread or sugar
and patting him on the neck. Does he lay his ears
back then? No! up they go, and the animal trys to
show in every way how pleased he is. I say I believe
horses are delighted to go out to work when they are
treated properly, but when they are not they try to
get away from the collar if possible, and will not let you
put the latter on at all ; in some cases it is a difficult
matter to put even the bridle on. My animals, when
they are loosed from the manger, turn round as much as
to say—"Harness me, I am pleased you are going to
take me out." I believe this is what they would say if
they could speak. Not only my own, but all horses
which are treated properly, look forward to going out
as much as a boy does being let loose from his studies.

One thing I have not dwelt very much upon in my
little work, that is—separating the dust from the hay or

chaff before the horse is fed upon it. This is most important. All the dust should be well sifted out before giving it to the animals. Dusty chaff ruins many horses. It is also injurious to give them flour of any kind in their food, viz., barley meal, pea meal, wheat meal, or, in fact, anything of that sort. When any of these meals are given to the horses they should always be mixed with damped chaff, so that the animals do not inhale the dust, if not, the latter clogs the horse's inside, and often is the cause of its being broken winded.

There is another thing I have not referred to, that is, when a horse has an accident and slips, or falls down, the driver occasionally gives the poor beast a kick in the ribs, or else a good lashing with the whip, to try and get him up. Now, to say the least of it, this is most inhuman, the horse would not fall down if he could help it. That is not all, however, when he is down he is very frightened and nervous, but when it comes to lashing him with the whip and kicking him, in less than three minutes the poor thing is all of a lather with fright. Any man who does this kind of thing ought to have six months' hard labour, yet it is done every day. I may say I am seldom in London but I see one or two horses down, and have helped to get many scores of them up on their feet again. While doing this one cannot help noticing what a difference there is where the driver is kind to his animal. When such is the case the horse is perfectly still until he is loosed. But on the other hand, when the animal has been used very badly, and the driver has been lashing him, the poor thing appears to struggle for dear life, and knocks himself about terribly. When a horse falls down, the first thing is to get some-one to keep its head down, as a rule it cannot hurt itself much if that is kept down. I need not say the harness should be loosed as much as possible, but, in cases where it cannot be unbuckled, it is better to cut a strap so that the whole harness can be cleared off. I might just mention here, many people get a little excited when trying to undo the traces, and so run a risk of getting kicked. Instead of doing that, the hames should be taken out of the collar, then they are already loosed. Only about three months before writing this chapter I saw a horse down in London, and the poor thing was hustled and thrashed to make it get up. This frightened it so much, it cut itself all to pieces. It would be well if people who have anything to do with horses would think a little before treating them so cruelly, it would be much better for the poor animals. Horses have feelings just the same as human beings, and though they cannot speak they know almost every word that is said to them. They understand when they are spoken to angrily, just the same as when they are spoken to kindly.

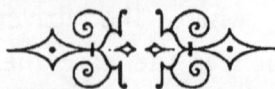

CHAPTER I.

FEEDING HORSES.

General Feeding—Nag Horses—Horses on a Farm—Waggon and Carmen's Horses—Judgment : when and how to use it, according to circumstances—Various systems of Feeding—Green Stuff—Dry Fodder—Preparation of Food.

THIS is a subject which is very important, and yet many people think anyone can feed a horse. Others again imagine a horse can be fed anyhow and upon anything, so long as it gets its food. Both these ideas are very wrong. Horses which have to work should be fed properly and on good food. It must be understood that whether they are thorough-bred carriage horses, hunters, or cart horses, it is the food which gives strength to the animals. The breed no doubt has a good deal to do with the spirit and pluck of a horse. At the same time no matter what breed the horse is, even if he is a thorough-bred, he can do but very little work unless kept up well with a good supply of nourishing food. The way in which a horse is fed, and what it is fed upon, is much more important than most people think.

Perhaps I may seem a little too particular on this point. I have never yet had a person in my employ that pleased me in feeding my horses, though I must not complain, as they usually look very well. When I used to feed them myself, I adopted rather a different plan of feeding them to many. Most people have their own way or idea of feeding horses. If they have been accustomed to a certain way it is very difficult to persuade them out of it, as it was the way their forefathers managed before them.

They say there is nothing new under the sun. There may be a good deal in that. Feeding horses properly is not new to many people. My father, thirty-four years ago, gave me a lesson on feeding horses, which I have never forgotten. I began to put into practice what he told me, and always found it answer well. Not only did it make the horses look well, but kept them in good condition. They were always ready for work when required.

For the benefit of any who may like to follow my method, I will mention what advice my father gave me, which has proved so helpful to me, and I am pleased to say many more as well. Only a few words : " Always feed horses so that they will lick the manager out clean, and give corn according to their work, and only give them a little at the time and often." These are the rules I have always carried out, and if I wanted to drive horses from thirty to sixty miles in a day, I have been able to do it without over-working my animals. One gentleman, who always carries out these simple rules for feeding his horses, I have known to drive 170 miles in two days, and at the end of the journey the horse he drove was not at all fatigued. I know this is a true statement, because I have ridden behind this horse myself, and it is still living, and a splendid goer at the age of 23.

Now, in treating upon the feeding of horses, it will be well to give several systems besides one which I have found to be the best for the animals, the most profitable to the owner, the way to get the most work out of the horse at the least expense, as well as keeping them in the best health, and making work a pleasure to them instead of a hardship. When horses are fed and treated properly, they appear to look forward with great pleasure to their feeding time ; and are almost as delighted to have their harness on and go to work. When they are fed badly, on inferior food, and treated unkindly, they seem to thoroughly understand it, and move as though they did not belong to anyone. Many of them show symptoms of rather not having a bridle put on than otherwise, and never seem to look forward with any degree of pleasure to being taken out of the stable. I believe in making animals happy, to a certain degree, as much as I do human beings. No animal appreciates the kindness bestowed upon it more than a horse, unless it is a dog. A horse which does a great deal of running, whether in harness or saddle, requires different treatment to a cart horse. I do not make so much difference between the two as some people, because any horse that works hard, though perhaps he only draws and does not trot, wants treating equally as well as one which runs ; only the latter requires more hay in proportion to the size of the animal than a horse which does the running, hay being to a horse what bread is to the human being. Of course, one which is required for great speed, should be fed in proportion to the amount of work it has to do on good nutritious food, both corn and hay of the best quality that can be obtained. It should always be of a dry nature, or at any rate the greater part of it, when speed is required, as the animal has to eat so much more green food than dry to get the nutriment required. A little green stuff is alright for horses which have to trot steadily. Take for instance a thorough-bred horse which has to run a race. The owner would not think of giving such an animal a bundle of green tares, clover, or even a feed on grass, when it is in training for the Derby or any other important race.

Horses which have to run very fast should have but very little green stuff, they will do much better without any at all before they have to run so fast. When they have so much green food the stomach expands very much. If a horse that is accustomed to running has been allowed to lie out to grass for a time, in a fortnight or three weeks the stomach will be from three to four inches larger round than it was before the animal was put out. I do not say this will be the case with every horse, especially where they have been kept well, but it will be found so in most cases. Now, it must not be supposed that I wish to rob the horses of their green food. I am in favour of them having it in the Spring and Summer, whenever it can be procured for them. Even carriage horses, which are used every day, if they are not driven too fast, are much better for a little green food once a day, as it keeps the body in such splendid condition. Properly speaking, during the summer months, green stuff is the natural food for horses. Take farmers' horses for instance, as soon as May comes in they live almost entirely on green food, especially in some parts of the country, yet they have to work very hard.

If they are allowed their corn just the same as when they had hay and straw cut up — especially if they have just a little chaff mixed with the corn,—farmers' horses become very fat while being fed on the green stuff, no matter whether it is tares, clover or grass. Their skin usually becomes as fine as a piece of silk, and the animals require no medicine at all during that time. While the horses are being fed in this way they should not be allowed to run very fast, if so, it is liable to injure their wind. Many people say they perspire very freely when they are fed upon green stuff. That is quite right, because when they are fed in this way the weather is usually hot, at the same time the green stuff may make a difference in this respect. When they do perspire it does not hurt them, it opens the pores of the skin, and in almost every case farmers' horses will put on flesh during the time they are fed on the greenmeat, and still do their usual work. In fact very often they do more.

Feeding Nag Horses.

Horses, like everything else, are very differently situated. Some horses, when they are allowed green stuff every day, look forward with such pleasure to it that they will not eat up their dry food, but wait anxiously for the green stuff. When this is the case it is far better to cut the green stuff in with the hay, especially where the person cuts their chaff every day.

This mixture must not be kept from day to day after it is cut up, if so, it will become heated and spoil the whole of it. If it is cut in the morning and is to be used during the day or night it is alright. I mention this because some people might be cutting chaff enough to last a week, and put green stuff in it. In that case it would do more harm than good. When horses have not a very good appetite during the summer, if they are working hard, a little green stuff cut up in the way described is very helpful to them. It does occur sometimes that horses go off their food, but it is very seldom if they are managed properly. I am often asked how much food a person should give a horse per day. This is really difficult to answer, for several reasons. To begin with, very often small horses will eat more than big ones. Then again, a great deal depends upon what work they do. Not only so, but the quality of the food makes a great difference in this respect. If it is good sound stuff they do not want so much of it as when it is poor stuff. I know some people say horses should have so many pounds of hay a day and so many pounds of corn. Now it is always unwise to give a stated quantity. Horses should be fed so that they clean it all up nicely. A spirited horse, which does but very little work, would soon go wrong if it were allowed the same amount of corn as when he was in full work.   Then again, when a horse is working very hard it cannot very well have too much to eat, at any rate, I never knew one that was really working very hard eat too much.

As regards the quantity, I repeat, that cannot be put into print for a person to go by.   I once had a little pony twelve hands high, which would eat a truss of hay, a bushel and a half of oats, and a bushel of bran in six days.   The oats weighed 42 lbs. to the bushel.   I also had a cob fourteen hands high.   A truss of hay lasted him a trifle over eight days, yet he was as fat and round as a mole and always ready for work,—proving the old saying true, " It is not what is eaten, but what is digested."   The cob, fourteen hands high, always used to eat his slowly and masticate it well, so that he used to get all the nutriment out of it but the other's food did not do him so much good, he, eating his quicker, and not masticating it so well.   I give this as an illustration of feeding horses.   The cob, fourteen hands high, had always been fed properly, and the pony had fallen into the hands of one or two people who were bad horse keepers before I had him.

Now, there is a great deal in feeding horses, and giving them as much as they will eat without wasting any.   When the person who attends to the horses sees any food at the bottom of the manger, he should give less food the next few days.   Whenever the horse does not clear the manger well out, give him less food each time, that is one of the best tonics a horse can have.   Of course there are times when a tonic does a horse a great deal of good, and the feeder should always observe whether the animal is perfectly well or not. If he is not, he wants nursing a little with something tempting, viz., a little scalded bran mixed up with some chaff and a very few oats, because sometimes a horse, when it has been on a journey, becomes faint.

Some horses will never clean their manger out well, they get so accustomed to the person who feeds them cleaning it out for them and throwing the remains on the ground, that they merely pick the best corn and chaff out and leave the other. I do not say if there is any rubbish at the bottom of the manger it ought not to be thrown out. That should all be swept out occasionally. When there is poor hay cut up among good, horses will often pick out the sweet and leave the mouldy. Now it cannot be expected that a horse is going to work hard on mouldy hay. Unless horses are properly fed, either the owner or the horses will have to suffer, most probably both will. I always believe in cutting the hay up for horses into chaff unless the animal occasionally gets off his feed, then he may have a little hay.

A horse perhaps, after a hard day's work, or a long journey, goes into the stable late at night, and the one who attends to it throws a lot of food into the manger, and the following morning they find a good part of it left. Now in such cases as these, it is always well to give the horse a nice little bit of sweet hay when leaving it at night, it tempts the horse to eat, and it will often eat the hay before it lies down. But as to the quantity of food that is put in the manger afterwards, the attendant must use his own judgment and give according to circumstances. When horses are working very hard, at either drawing or running, there is nothing like changing their corn occasionally, giving them a few split peas or beans in addition to the oats. Some people do not think much of oats for horses, but say maize, barley and French buckwheat should all be split up together with the beans and given to them. My experience is that horses which have been long journeys, especially if they have been running hard, do best on oats, with a few split peas or beans mixed with them for a change. It is a capital plan to have an oat crusher, as in some instances the oats pass through undigested, because some horses bolt a great many of the oats down, that is to say they swallow them without masticating them ; but it is not a good plan to buy them crushed, if so they should always be bought by weight, as they are very often inferior oats crushed for the purpose. I always like to see the corn whole and sound.

I know some Veterinary Surgeons go against crushed oats, but I merely speak as I find. I had a great deal to do with cart horses in my younger days, being brought up on a farm. Then later on, I had to do with hunters and carriage horses. In fact, I may say I have been among horses ever since I was six years old, so that I think I may speak from experience. If ever I hear of a horse doing extra work, or going on a long journey and standing it without fatigue, my first enquiry is, "How is the horse fed, and what corn is used ? " In almost every case the principal part of the food is oats. What do our sportsmen use for their horses which they hunt with from three to four times a week ? Why, the best oats they can buy, and but very little else. I think it is not important for me to state in this little work how to feed hunters, because most of those who hunt usually have a reliable man to see after them, who is thoroughly up to his work.

Though there are plenty who keep horses, and may either feed them themselves or have a man to do it, I do not care who the man is, if he is not fond of horses he is not a good horse-keeper. A person may be fond of horses, and not know how to feed them, but on the other hand, if he is not fond of the animals he will never know how to feed them properly, because he will not take pains enough. Now the London cab horses, which work very hard and make long days, will often eat half-a-bushel of oats in a day, and yet are not big horses, but run from fourteen hands, two inches to fifteen hands, three inches. It is very seldom the poorest cab master allows his horses less than a peck and a half a day. Very often if carriage-horses about that size were to be allowed as much corn there would be no holding them. I consider a peck a day, with good hay, enough to feed a horse say from fourteen to fifteen and a half hands, that is of course one which only does just medium work. Should a carriage horse go a long journey and have extra work to do, by all means give it extra corn, as much as it will eat.

Those who drive can soon tell whether the horse is having too much corn in proportion to the amount of work it is doing. If it is. the legs will soon begin to swell, and unless it is very quiet in harness will show symptoms of irritability. I always like a horse to have a little bran, and for the last twenty years whatever horses I have had under my care, I have allowed at least a peck of bran a week, sometimes a bushel. No stable should be without bran in it. As regards the quantity of hay to be given or cut up into chaff, the attendant must go according to his judgment, so that they clear it up and not waste it. If a person only keeps one horse he should have a small chaff cutter, and cut his own hay, as the chaff supplied by corn chandlers is frequently one half of it very inferior stuff, as they get a little over-heated hay that smells strongly to scent the whole and then cut nasty mouldy hay with it, and thousands of poor horses have to suffer through this.

In no case should too much be put in the manger, if so the horse does not relish his food, neither does it do him so much good. He throws a great deal of it out, and in many cases a good deal of it is wasted. There is a difficulty when a horse comes in late at night : he must have a good supper, but as I have hitherto stated, if a little bit of hay is put in the rack the horse will do much better, will be more lively in the morning, and quite ready for breakfast. All chaff which is likely to have any dust in it should be well sifted before it is put in the manger, because when there is dust in chaff it is very injurious to the horses. A machine is now in use which separates all dust from the hay, so as to avoid this being mixed with the chaff after it is cut. The feeder can always tell whether there is dust in the chaff or not, as the horse will leave it at the bottom of the manger. When a horse is going on a journey, either saddle or in harness, and is going to run at the rate of eight or more miles an hour first thing in the morning, the animal should finish its food half-an-hour at least before it starts. This makes a great deal of difference to it. Many horses are driven too fast, and it injures the animal's wind a good deal, in fact, there are more horses have their wind broken in this way than from any other cause. If a horse's stomach is full it should be let go steadily for the first two miles, and allowed to walk up all the hills. When care is taken in this respect it does not hurt the horse's wind. But when on a long journey, the horse should never be driven more than from eight to twelve miles, without having a little water and a mouthful of hay. My plan is to take a small nose bag when on a long journey, and about every eight or ten miles give the horse water with a mouthful or two of food. I do not mean to say I put my horse in the stable every eight or ten miles, but just put the nose-bag on and let him have a little as he stands.

People injure their horses by driving from twenty to thirty miles without giving the poor beasts either a bit of food or drop of water. I differ from the majority of people as regards watering horses, but as this is an important matter I am giving it a small chapter to itself. A horse which does a great deal of fast running should always have the best hay cut up into chaff, and nothing but the best corn given it, as such an animal does not want blowing out with a lot of hay. A horse which trots steadily should have

more hay and less corn, as I said before, be fed according to
the work. Corn of course has more stay in it than hay.
Take a horse which goes hunting three or four times a week :
hay by itself is scarcely any use to such an animal, because
it often goes so long without food before getting home.
Sometimes it does not get more than a mouthful of water
and a little hay until it gets in the stable late at night,
therefore it requires plenty of the best corn. Hunters
should really be turned out three months every Summer ;
then they would last as long as horses, of the same
stamp and stamina, that are working in an ordinary way
all the year round.

Hunting is very trying work for a horse, though the
animals may be as fond of it as the riders themselves.
There is many an old hunter that is put on one side as
worn out, and has to go on farm work, especially those
which will draw. Some will not, after they have been
accustomed to hunting and saddle back, but when these
horses which will work are drawing the plough, if the fox
hounds happen to go past where they are, very often they
will break loose from the team and run after the hounds.
I mention this, because some people say it is so cruel for
horses to have to run across ploughed fields and jump
hedges and gates, but when they are trained to it they look
forward to it and enjoy it as much as they do eating their
food. It is well not to give hunters clover, and but very
few beans. They should have good oats and the best
meadow hay. A few dried tares cut into chaff is a good
thing for them. I mention this to avoid giving them clover

and beans, because they are not so good for the wind as oats and meadow hay.

If one horse is fed on clover and a large quantity of beans, and another on good meadow hay and oats, it will be noticed at once that the wind of the horse fed on the latter stuff will be much better than the other. I draw attention to this fact, because earlier in this Chapter I recommend beans to be given when they are working hard, and a novice might think he should give a hunter a good quantity of beans, and in some cases if this were done it would be fatal. I never allow more than five or six pounds of beans a day to a carriage horse, but ten pounds a day would not hurt a draught horse, that is if he is worked hard.

There is one other class of horse keepers I should like to mention as regards feeding. It was thinking of this latter class which first induced me to write a book on the horse. To those who have not been accustomed to horses a few hints may be useful. I have seen so many horses in my travels mismanaged and badly fed ; sometimes they are,—what shall I say ?—killed by kindness. Many people when they have a horse think they should wet their food and give it to them what they call sloppy, as they think a horse has so much less trouble to eat it. This is a great mistake. If the food is made wet in the morning, very often the horse swallows it down much quicker and does not masticate it properly, therefore a great deal of good food is spoiled in this way. For instance—costermongers, greengrocers, and such as these, will often get a little bran, chaff and oats, and mix them altogether into a sloppy mess, and the horse or donkey, whichever the case may be, swallows it down quickly and does not masticate half of it. A certain juice and saliva which should mix with the food does not do so, and when a horse is fed with this sloppy food it nearly always causes the bowels to become very much relaxed. Nothing will upset horses quicker than giving them a lot of sloppy food.

Some people may argue in this way, and say that when they drink water it mixes with the food, but that is another thing altogether. It is natural for a horse to masticate his food well. Nature has supplied these noble animals with grinders sufficient for masticating any corn or hay which is grown. Of course, when a horse is old and its teeth are worn out, that is a different matter, then it should have its corn boiled, but not water put upon the chaff, only the boiled damp corn mixed with the chaff. They are able in this way to get the whole of the nutriment out of the boiled corn, because it is already soft and wants but very little mastication. A great deal more nutriment is got out of the chaff too when it is dry, or partly so, than when it is mixed with water. If the chaff and corn are both made sloppy you will be almost sure to see the hair stand the wrong way upon the body, and the ribs very close to the skin. The latter seems to fit him tightly, like a kid glove on a person's hand. When the skin is tight on a horse in this way and will not move easily, it is a sign the animal is doing badly.

If a horse is very weak or old, there is nothing like stewing a little linseed and mixing it with a little chaff and bran and giving it to the horse, but not too much at a time to make it sloppy. This is very nutritious, and helps the animal very much. The linseed should be stewed so that it is very thick and glutinous.

Horses will always do well on this, but they ought not to be worked hard or excited when they have it, if so, it is likely to relax the bowels a little too much. When a horse is very costive a better thing could not be given, if a little bran is used with it. Some horses are naturally very costive, so much so that I have seen them pass blood with their motions. When this is the case the animals should always have a little bran given them in every meal. A nervous horse is just the opposite and is almost sure to become relaxed in the bowels.

Waggon and Carmen's Horses.

I do not think it is necessary for me to treat largely on feeding the heavy horses, that are used principally upon the roads, no matter whether they are brewers', millers', merchants' or ordinary carmen's horses, they usually fare better than farmers' horses in a general way, for the simple reason, as a rule they carry a nose bag with them. It does not hurt the horses to work hard if they are fed often. A horse has a much smaller stomach than any other animal, that is according to its size.

It is said when a horse lies out to grass it usually eats twenty hours out of the twenty-four, but this is exaggerating it a little. They often eat more than twelve hours out of twenty-four, and when the grass is very short they may be nibbling eighteen hours out of twenty-four. Now carmen, if they are only going to stop five minutes, should always hang the nose bag on their horses for two reasons—First, because the anmials keep in much better condition, and do not become faint in their work when they feed often; secondly, when they have their nose bags on it teaches them to stand much quieter, and they can be trusted better. Some carmen make a mistake in buckling the nose bag on tightly. The nose should not be pressed into the food when it is buckled on, if so, they breathe into it so much. Their head should be at liberty, so that they can place their nose bag on the ground, but in all cases great care should be taken as regards buckling their bridles on tightly, because when a horse holds his head down, if the bridle does not fit well, occasionally the animal gets it off. Some horses will run away if their bridle comes off. It is far better to take care in this respect, and be on the safe side. If I kept draught horses, and they travelled by road, I should have a small table constructed to fold up like a dandy chair, and carry it under the cart or waggon for the nose bag to rest upon while the horse is eating. It may often be noticed that when a horse has his nose bag on and he cannot reach the food, he will keep tossing his head up to get at it, and by this means a great deal of the corn falls out upon the ground. There are hundreds of sacks of corn wasted in this way throughout the country, when it might easily be saved by giving the horse something to rest the nose bag on.   A table to stand could easily be made as I have described.   The legs should have a hinge on the top and a hook on the opposite side, so that when the hook is undone the table legs will fold underneath, and it can be hung on somewhere under the cart or waggon when it is not in use.

Feeding Horses on the Farm.

Some who read this little work may think, surely farmers know how to feed their horses without being obliged to read books.   No doubt there are many farmers who do know how to feed horses, but I am sorry to say there are some who do not understand them, or in other words, do not take the trouble over their horses in this respect that they ought.   Farmers' horses ought not to be fed anyhow when they are worked hard at different periods of the year, especially in the autumn.

There is a great amount of good food wasted on the farm through the carelessness of the horse keeper.   Farmers are the people who ought to use their horses well and keep them in splendid condition, for two reasons.   First, they grow all the food the horses are fed upon, or at any rate the greater part of it ;   secondly, most farmers do a little breeding in young horses, if not, they ought to do so, if possible, then they usually have one or two for sale.   The better condition they are kept in, naturally the more money they make.   That is not all however, when a horse is kept in prime condition it really eats less food.   When horses are poor, and what some people call half-starved, the food does not appear to do them half as much good. It is therefore much better to keep horses in good working order and their ribs well covered with fat. During the long days of Summer I have been very much surprised in going through farmyards to see what a great waste there is with the green stuff, viz., clover and tares. It is usually carried with a fork from the cart to the stable, and little pieces drop on the way, which are wasted. I consider on a farm where there are from seven to twelve horses kept, in almost every case there is nearly a load of hay lost. That is to say, if more care were taken in feeding the animals, and instead of dropping so much about, it was cut and dried, it would make a load of hay, which some years would be worth five pounds, but at any time three pounds.

Farmers are usually careful with their money, but not with their money's worth in these little things. It is just the same in feeding horses in the stable. Many farmers, especially those in the midland counties, will rack their horses up the last thing at night with a bundle of hay, giving a little piece to each horse in the rack above the manger, usually throwing it in with the fork, so that the horses can pull it out as they like. When it is meadow hay, and the horsekeeper takes it from where the bundle lies in the stable, right up between the horses, on the fork, it drops all the way. Sometimes, where the hay is very short, it almost covers the ground. Where the attendant has been carrying it along from horse to horse, on a fork, I have known from two to three pounds of hay wasted out of about thirty-six pounds which has been given the horses in this way.

This is of course a great waste, and yet I have never known the farmer complain to the horse keeper. He usually takes it as the natural course of things that he has always been accustomed to. Now, suppose this is done for say seven months in the year, and from thirty-six to fifty pounds of hay is used each night, two pounds out of that amount is wasted every day, that means over seven trusses of hay wasted in two hundred and ten days. I am very much under-rating it by these rough figures, as, when horses are racked up with hay, unless it is given very carefully, and is of the best quality, there is much wasted. The animals get it down under their feet and trample on it. Many farmers do not give their cart horses any hay cut up into chaff, but merely cut up wheat, barley, and oat straw, not that they always mix it together, but oat and wheat straw are usually what they use, the latter in preference. When they have thrashed the wheat they use the chaff which comes from the ear.

It is most remarkable what good condition horses will keep in if fed and managed properly. You might go to two farmers, and find one allows his horses more corn than his neighbour, and yet they are not in anything like so good condition as the other horses. This may appear very strange to many readers. A good horse keeper always gives his horses, but a very little at a time, and very seldom takes them out in the morning unless he gives them two hours' bait, as they call it. That is, feeding them two hours before they begin to work. When they are fed in this way the horses enjoy their food and masticate it well.

Generally speaking, the horses should not have any more food till they have cleared their manger out clean. The attendant should see to it that all the dust is out of the chaff, as the horses object to this being in their food very much. Bran, oats, and beans are usually the farmers' bill of fare in the way of corn for horses in many parts of the country. A farmer who understands his business well, as a rule, allows his horses a peck a day, and a little bran in addition to that. It must be understood that although the work a cart horse does is hard, yet it is steady. A person who understands feeding his horses properly will be able to get as much work out of them on straw cut up and wheat chaff as another will who feeds on half hay cut up and half straw, and yet both are allowed the same quantity of corn. Horses always enjoy their food so much more if they only have a little at a time, no matter whether they are cart or carriage horses. When too much food is put in the manger they often throw some out of the sides. That is not the worst part of it however. When they have more food given them than they can eat, especially cut stuff, such as chaff or corn, they blow in it. It must be understood that a horse does not eat in the same way as a cow. When the latter is feeding on chaff or anything similar to that, she pushes her mouth down and licks the food in with her tongue, fills her mouth, and then holds her head up until she finishes chewing it. Not so with horses, they only take a little at a time in the mouth, and usually keep the nose in the manger the whole of the time they are eating. They keep turning the food over and over, picking the sweetest out first, consequently they

c

breathe upon it. When there is a large quantity in the manger it becomes warm and moist with their breath, and the horses very much object to this. I suppose no animal is more particular in this respect than the horse. When they are only given a little at a time it prevents this, and so saves a great deal of waste. I have known good horse keepers give their horses from eight to twelve feeds, or in other words divide their morning meal into eight or twelve portions, giving a little at the time before taking them out at six o'clock in the morning, and the smaller lots they have the better they seem to get on. In some parts of the country they work from six to two, without having a bit of food or water during that time. I consider this a cruel practice, and one that should be avoided by taking a nose bag of food so that they can have a little at intervals as they work. Then when they get home the horsekeeper commences again giving a small feed at a time the whole of the afternoon, usually up to six or seven o'clock, and sometimes it is nine o'clock before they get their last feed. This may seem ridiculous to those people who throw their food in the manger and only give them one large feed before they take them out to work. Notwithstanding this, it is the proper way to feed them where there is a number of horses kept, especially if the nose bag is not used during the day. When farmers give their horses hay it is better to cut it up with the straw into chaff, the horses relish it so much better and seem to clear the chaff up. Whatever food the horses may be fed on, it is best to divide it and give a little at a time, unless it is green stuff—that alters the case, and even then too much should not be given them at a time, because the horses relish it so much better if it is given them fresh.

Perhaps it will be as well to remark here that horses used by merchants, millers, brewers, and for other waggon purposes should not be subjected to the treatment described in this chapter.

A few years ago farmers were more accustomed to give their horses loose hay than they are at the present day. They seldom cut anything up, but this idea of feeding is fast dying out. When the horse had corn it used to be put in the manger without anything else. When this is the case the horses eat it up greedily and do not masticate it properly. Consequently, much of it passes through them not digested, and the horses lose the nutriment of the corn, and so get robbed of its strength, and not only so, but when everything is cut it is much more economical. If farmers feed their horses on hay they must be at a loss to a certain extent through waste, as the animals are bound to pull it under their feet and trample it in the manure.

Where there are twelve horses kept on a farm, and they are fed in the old fashioned way, viz., on the hay and loose corn thrown in the manger, there is a sufficient quantity of food wasted for two horses, so that with careful feeding, fourteen horses might be fed and kept on the same amount of food that twelve are. And it is not only the waste, bu the horses do not look so well, nor are they able to do the work they could if their fodder is cut up for them.

C2

I do not think it is worth while wasting space
to tell the farmers what to give their horses, as most
of them are brought up to their business, or at any
rate should be.   There are two things, however, which
I think it will not be out of place for me to mention.
These are—when they sow the wheat in the autumn,
the horses have to work harder than any other part
of the season, many of the farmers do not give their
horses any extra corn, consequently they go down in
condition, and are not able to pick up again till the
following spring.   This is a great drawback to many
farmers.

When the horses are working hard during the
autumn they should have at least half as much again
corn and good hay, as horses become faint during
the autumn, the most trying time for horses, because
they are changing their coats as well as working so
hard.   I have known farmers at that time of the year
work their horses for nine hours, and never give them
food or water.   This is what may be termed cruelty to
animals, of course it is done without a thought.   I
know it is done perhaps more in Bedfordshire,
Cambridgeshire and Lincolnshire, and several other
counties.   If farmers would only go to the expense
of a nose bag, while the men are having their lunch
the horses could have a feed as well.   If this is done
they would do their work much easier.   I am
pleased to say this is carried out in many parts of
the country, especially in the South and the extreme

North, and the men have their lunch while the horses feed at the same time. This is how it should be.

I give this last page on the feeding of farmers' horses, hoping that those who have not used the nose bag may take the hint. Farmers should always take care and have a few old split beans or peas by them when their horses are working very hard. There is more nutriment in beans, peas, and tares, than any other corn, and as the farmers are able to grow all these they can afford to use them for their cattle. When horses are well fed by a farmer his money is not spent badly, as they last longer, and do the work easier, not only so, but if he wants to sell the horses they will make a far longer price ; so it is better all round.

Many farmers when they begin to give their horses green stuff knock off the corn. Now this is wrong. When horses are put on green stuff, and have been accustomed to have corn, if the latter is taken off all at once the bowels become very much relaxed, and in cases where the horses are worked hard they often have the gripes. A horse which is worked hard, even if he does have green stuff, should always have a little corn. This will keep them in splendid condition. It is penny wise and pound foolish, when the corn is taken off altogether, as by doing this the farmer is often money out of pocket. What I mean by that is, the horse will often be laid up, and that is loss of time with their work, which is very important during the summer months when farmers are busy. If farmers cannot afford to keep their horses well, they should not keep so many of them.   It is a certain fact if they will not pay for keeping well they will not pay for keeping badly.   And if a horse is not worth its food it ought not to be kept.

CHAPTER II.

WATERING HORSES.

Experiences—Bad watering and its results—Methods of giving water—
Let the horses drink as they like.

I AM afraid my readers will think I am rather bigoted on
this subject, viz., watering the horse. I will only write
what I have found to be right from my own practical
experience, and give the reason why I think my way better
than that of most people. Most people are very careful
not to let a horse have too much water to drink at a time,
and during the very cold weather they do not allow it to
drink any cold water. This practice is all very well if the
owner always attends to his own horses, or even if the
attendant has them under his own control, but when a
horse which has been accustomed to have chilled water in
cold weather is put up at a livery stable, and gets at a
pailful of water, he soon empties it, and the result is, colic or
gripes. I have known many cases where a horse has not
been accustomed to cold water, when it has been given a
pail full, it has brought on the staggers almost at once,
and in some cases the animal has lost its life. Then there

are others who only allow their horses but a very little to drink at a time and never chill the water at all.   There are great difficulties attending this practice also.   For instance, we will say here is a horse which is never allowed to drink when he is hot ; he goes on a long journey, and the ostler at the hotel, or the place where the animal is put up, happens to turn his head for a minute, or it may be while he is washing the horse's legs, it gets at the pail of water he is washing the legs with, or manages to get to a trough of water ; in this case the probabilities are, it either brings on gripes or causes the bowels to be very much relaxed.

My system in watering horses is to let them drink what they want, and never allow them to go too long between the periods of drinking.   Suppose a person has two horses, and as a rule they drink about the same quantity of water, one of them is allowed to have as much as it likes, and the other has the pail taken away from it before it has had sufficient, the owner will find that the one which has enough and to spare will not drink anything like so much in a week as the other one which has barely sufficient. Some may ask, "Why is this?"   It is easily answered in a few words.   A horse is a very intelligent animal, and when it is drinking, if it is accustomed to have the pail taken away before it has had sufficient, it will drink much faster and try and keep its head in the water as long as possible, so as to prevent the attendant taking the pail away.   When I am going on a long journey with a horse, no matter whether it is hot or cold, I try it with water every eight or twelve miles, and always allow the animal to drink as much as it wants.

I have had many horses under my care ever since I was fourteen years old, and have never had one with the gripes. Many people have lost their horses through this affection When such is the case, I often put it down to not allowing the horses to drink what they like, and keeping them too long between the periods of drinking. I always try my horses with water the last thing at night and the first thing in the morning, besides two and three times during the day, even if they are only in the stable and doing no work. I recommend this system because I have found it answer so well. If I am driving in the country and there is a pond on the side of the road I let my horses drink what they want. When I am boarding at any hotel and the ostler comes up to take charge of my horse, I always say, " Give him some water," and the answer I get is, " When he gets cool, sir, not before, as it will give him the gripes," then I say, " Give it to him at once, I will take all the responsibility." I am usually inquisitive enough to ask why they do not give the horses water when they come in ; in almost every case the reply is, " We never let them drink when they are hot if we can help it." " But do not the horses get hold of the pail sometimes when you are looking away or speaking to someone ? " " They are almost sure to have the gripes if they do, sir, and that makes us a bit careful." These are the answers I usually get from the ostlers and those who have to attend to the horses. It is quite right for them to be careful, because they have a great many horses come into their hands, and not two out of every hundred are allowed as much water as they like, especially when they are hot. With my system, the

horse is safe either way, whether it has water at once
or waits.   Many of the ostlers have told me that if their
employers knew they gave a horse water when it was hot
they would be discharged at once.

If my horses come home the coldest night in Winter, or the
hottest day in Summer, they are always allowed to drink as
much water as they like.   Some people have a pail of water
fastened in the manger for them to drink when they like,
others again will have an iron manger and a place arranged
purposely to keep water in, with a plug at the bottom, so
that it can easily be washed out.   I do not like this method
so well, and I will give my reasons why.   When the mangers
are arranged in this way, horses get accustomed to keep
putting their mouth in the food and then in the water, and
by doing this they often make their food quite wet.   When
the food is moistened the horse swallows it much quicker
than it otherwise would, and does not masticate it as it
ought.   This is the principal reason why I say horses
should not have water standing by them the whole of the
time they are eating.   There are cases where a little time
should be spent in allowing a horse to drink as much
water as it likes.   Suppose, for instance, the owner or
attendant has been driving very quickly and the animal has
been on a long journey, and has not been offered any water
for two hours.   If a horse in such a state as this were
allowed to drink as much as it liked it might do the animal
injury, so that a little care and judgment should be
used in this respect.   I have never had a horse of mine
meet with this misfortune, because when I am on a journey

I always see my animals are attended to properly, both as regards feeding or watering, before I have anything myself. When I have driven a distance of over ten miles I never leave my horses to any ostler, and when they have to be put up at a strange place for the night I do not leave the stable until all is made comfortable and the animals are bedded down to my satisfaction. If those who travel long distances with their horses would only take this precaution, and see that everything is comfortable for them, and they have good food and water, and a nice clean bed, they would find heir horses as fresh as possible the next morning.

Farm horses want treating a little differently. When they have been worked very hard for seven or eight hours I do not say it would be very wise to allow them to drink as much as they liked directly they come in, unless they have been accustomed to it. When a horse will not drink before he goes out in the morning, and works hard all day without any water, naturally he becomes very thirsty, and if he were allowed as much water as he liked to drink, probably the result would be an attack of the gripes. In such cases as this the horse keeper or attendant must be guided, to a certain extent, by circumstances. I feel confident that those who try the experiment of giving horses water frequently, whether they are working or standing in the stable will find it answer well, especially when the animal is working. When working hard they need almost double the quantity of water they require when lying down in a straw yard or loose box. It is not my intention to give the quantity of water a horse should have because they vary in this respect as much as in the food they consume, but I simply advise those who have horses to let them have water as they like, and as often as they will drink it.   If this rule were carried out better there would be less work for the veterinary surgeons and far less trouble and disappointment to the owners of the horses.

CHAPTER III.

TYING HORSES UP IN THE STABLE.

Horses should be tied up—How to fasten them properly—Tricks and misdoings of Horses if let loose.

SOME people object to horses being tied up; but this cannot be very well avoided, because in a large stable, or even where one has only got three or four horses, they cannot always afford room to give them a loose box. It is therefore wrong to say horses should not be tied up. Now it is rather a difficult matter to say how horses should be tied up, or what to use for the purpose. A horse which is fed well, and does but little work, becomes very restless in the stable, and very soon gets in the way of working its head up and down so as to make a noise with the ring and plug, especially where iron mangers are used.

Years ago it was very seldom one went into a stable and saw anything but a rope leader, or what is called in some parts of the country a halter, or chap rein. This had a hole at one end, so that it could be fastened with a slip knot to the head stall, then there was an ordinary wooden plug to keep it weighed down. There should be

a ring attached to the manger. When the halter is tied
to the ring without the plug, and the horse goes to scratch
its mane or neck, it gets the halter underneath the heel of
its hind foot and throws itself down.   In some cases a
horse has been known to break its neck in this way, or
rupture its neck, and sometimes never recover.

It may seem ridiculous on my part to mention this,
as many of my readers have always been accustomed to
horses since their boyhood, and will say at once, "Anyone
knows a plug or weight must be put on the end of the
halter, which fastens them to the manger." Such people
as these must forgive me for bringing forward such a simple
thing as this, but it must be remembered there are so
many fresh people buy horses, both for work and pleasure,
who know nothing whatever of the habits of the animals,
or how they should be tied up in the stable—and some of
those who do know about it seem never to realise the
importance of carefulness in these little matters—or what
serious consequences may arise if these are neglected.
I have come across serious accidents on account of people
tying the horses up without a plug on the end of their
halter in the stable.   A horse should not be left one hour
with its harness off, unless there is a plug or weight on the
end of the tether.   Many people during the last few years
have used a chain in preference to the rope, as it lasts
much longer, but where this is used it makes a great noise,
and if the stable happens to be near the dwelling house it
is very annoying.

A better plan than using the chain is to have a strong strap, rather narrow, with a wooden plug on the end, weighing at the least from 1lb. to 1½lbs., with about a foot or a foot and a half of chain at the top, which can be fastened to the head stall that the horse wears in the stable. If managed in this way the halter does not make any noise working up and down in the ring. When the chain is used, and makes so much noise, the horse gets accustomed to working it for pastime. It is always best to have a leather head stall to tie the horse up with, so that it can be buckled underneath the throat, if not, old horses will sometimes slip a hemp halter.

I may mention here that I am referring more to farmers and carmen, who often tie their horses up with a hemp halter. Though there may not be more than from two to seven horses in the stable, if one, unfortunately, manages to slip its halter and get amongst the others, it often brings about unpleasant results, sometimes one of the horses having a kick, or meeting with an accident of some kind, which means a veterinary surgeon's bill.

I have known many cases where horses have slipped their halters, and got loose in the night when the farmers have used the ordinary hemp halter. It is generally a case of " Penny wise and pound foolish." When a leather head stall is once bought it lasts for years.

Then again, when a number of horses are kept in one stable, the man who has charge of them should always see that the head stall and chain, or whatever they are tied up to the manger with, is not likely to be broken, or worn through, so as to avoid accidents ; because when horses once get loose they are almost sure to go and ask their companions how they are.   If they do not do this they find their way to the corner of the stable where the food is kept, and their neighbours seem to quite understand what they are doing, and become restless and very often commence kicking.   Many accidents arise through neglecting to see to these little things, so I would say to all who have charge of horses, be sure and see they are tied up securely.

CHAPTER IV.

BEDDING DOWN IN THE STABLE.

There are various ways of bedding down horses in the stable: some are better than others.

I HAVE spent much of my time in stables of all kinds, even in the mines and coal pits; and all kinds of places where a number of horses are kept have come under my notice. I was asked a few years ago by one of the proprietors of the *Live Stock Journal* to write articles for that paper on the management of the horse while in the stable. Being very busy at the time with several other papers, on quite a different subject to the horse, I could not spare time just then to comply with the request.

I always say more horses are lamed while in the stable than when at work. This is my opinion. Take for instance a gentleman's horse, no matter whether it is a hackney carriage horse, hunter, or pony, frequently we find it gets lamed from some cause or another. I will mention some things, which I believe have a great deal to do with horses becoming lame.

Most gentlemen are very particular about having the stable floor cleaned, and will even have the place washed under the horses' feet. Some have this done once a day, while others are content with having it done once a week. Now many of these stable floors are made of hard Staffordshire blue bricks, or sometimes asphalte, and by washing them down so often it makes the surface very smooth. Of course the bottoms of the stables are not always made of the two things I have named, occasionally cement is used, and I will explain further on how much better this is for the horses' feet.

In my younger days I was always very much surprised to see the number of horses that were lame. They were mostly carriage horses. If you took one hundred of the latter, and one hundred waggon and cart horses, including those belonging to farmers in the surrounding districts, you would find there would be from fifteen to thirty of the carriage horses laid up at certain periods of the year, and you would only find from three to seven cart horses lame out of the whole hundred. I could not help feeling sure in my own mind (although young at the time) there was some cause for this, and as my nature has always been a rather inquisitive one, I made many enquiries from different gentlemen, coachmen, grooms, carmen, and farmers of my acquaintance, as to the reason of it. I did not get a satisfactory answer to my mind from any of them.

The farmers said, " Ah, you see we do not pamper our horses like gentlemen do." The carmen's answer was, " Carriage horses are better bred than cart horses, and are more likely to go wrong." Coachmen and grooms usually said, "Horse dealers are such rogues, they take our masters in with horses that have been pampered up, and they go lame after we begin to work them." No doubt there is a great deal in the latter, and I am sorry to say it is often the case. I asked many gentlemen how they accounted for the large percentage of carriage horses and trotting nags going lame, so much more than draught horses, (which usually go by the name of cart horses). All I could get from that quarter was, "Oh, that has always been a mystery, which has not been fathomed yet."

None of these answers seemed to be satisfactory, and I thought I would at least try to find out why there was such a difference between the carriage and cart horses in this respect, and so clear the mystery up.

In my boyhood I have watched horses in the stables for hours, with just a little glimmer of light. In the Winter evenings I would rather spend two or three hours lying upon a piece of straw in the stable, watching the habits of the horses, than I would by the fireside. Many things I learned during these times of observation have been very useful to me during my life, but I cannot say that I have solved the mystery yet. In many instances the cause cannot be accounted for.

Some fourteen or fifteen years ago I discovered a clue to it, and thought I had found out the secret at last ; but although it may not always be the cause of the horses going lame, I believe, at any rate, it is in many

instances.   I noticed that when the brick or asphalte
floors were washed down, it naturally made them very
slippery, and when the horses wanted to lie down they
managed that very well.   The getting up part, however,
was quite another matter, as their feet often slipped
from under them.   I have seen many horses try to rise
from the bricks, and fall right down again.

It must be understood that the horse is quite different
from the cow.   The latter goes down upon its knees first,
and then drops down altogether.   The horse drops his
hind part first, and when getting up puts his fore feet out
first ; while the cow rises with its hind legs first, and gets up
very steadily on to its fore feet—there is no slipping with the
cow.   When, however, horses make a start to get up, they
spring upon their legs, drawing their feet up, and if the
floor of the stable, where they are lying, is smooth, they
often make a slip, even if they do not go right down.

To prove this statement, let those who may be going
through London streets, where there are asphalte roads,
notice horses which have fallen down, what a difficult
matter it is for them to get upon their feet again.   In many
cases the drivers are unable to get them up till they
fetch some sand, and spread round where the horse falls.
Now a horse standing in a stable, paved as many of
them are at the present day, is in just about the same
position.

Those who are interested in, and watch the noble
animals in the stable, will notice that seventeen out of

every twenty, especially carriage horses and such like, will paw and rake the straw from underneath them with their fore feet, and where this is the case their knees have to go on the bare bricks, or hard floor. When they go to rise, they spread their fore feet right out from under them, and many of them jump up quickly, and cannot get foothold. It is not a matter of ricking themselves in their fore feet so much as in their hinder parts, especially the round bone, although occasionally they do rick their shoulders.

The owners, no doubt, think they are doing the animals a kindness when they have the stable floor well washed down. To my mind it is a great mistake to wash where the horse stands ; if it must be done, it is best to cover it up with either chaff, fine sand, moss peat, or even dry earth put down would be better than leaving it bare. Of course where there are drains in stables, to carry the urine away, they must be washed down and cleaned occasionally, if not, the stable smells very badly.

Many gentlemen are very much against using short straw for bedding horses down in the stable. When buying they will have the long fine wheat straw. This, I think, is a great mistake, the longer it is the worse it is for the horses' knees. I will explain why this is so. Before horses lie down, as I have said before, in seventeen cases out of every twenty they paw with the front feet on the ground, and pull the straw back from underneath them. It not only makes them slip when they go to get up, but their knees come into contact with the hard floor, so much so that horses often knock the hair off the knees when in the stable. I have known horses have corns on their knees through fetching their bedding from under them where the stable has been washed down clean every day ; in some cases the hair has grown long and coarse, and has given the horse the appearance of having been down and cut its knees. This, of course, would spoil the sale of the horse.

Sometimes there is a little chaff that the horse is continually throwing out of the manger, this, with dust and chaff from the wheat straw, together with the dampness of the floor, in a very few weeks forms into quite a cake on the top of the floor, when the latter is not washed down. This enables the horse to get good foothold when it attempts to get up, even if it has scraped the straw from under its feet. Farmers' stables, as a rule, are only cleaned out and swept, not washed at all, except in a very few instances, and in these only at the back, where the horse's hind quarters come, and, therefore, when the horse attempts to get up, he has a good foothold, and so has no difficulty in getting on his feet. The same thing may be said of carmen's and all cart horses.

I do not think one cart horse in a thousand, where the stables are only swept out and the horses are enabled to get a good foothold, become lame through lying down in the stable, or injure themselves in any way. The loose stuff on the floor fills the crevices up, if there are any.

I find that where horses are bedded down on moss peat there are very few cases of lameness, though I know many people object to it, as it has a dirty appearance in the stable. I always use it myself, though I do not keep many horses, but I find they do much better on moss peat than they do on straw, and it is considerably cheaper. Certainly it may make the stables look a little more dirty, but it has several advantages over the straw.

In the first place, if the horse scratches, or paws about with his fore feet, it does not pull it all in a heap the same as it does the straw, as there still remains some of it on the floor, which keeps the horses' knees from the hard bottom. That is a great thing in my opinion, and at the same time the horse has got good foothold to get up again.

Another objection many people have to moss peat is, that it stops the drains. Now may I just mention that where this is in use, no drains are required, and they ought to be stopped up. Moss peat is a deodorizer, and absorbs all the urine, and therefore holds all the ammonia and strong smell. Wherever the urine is, just that piece should be taken up, and some dry stuff thrown in its place. I never found any of the urine went into the drains where a proper quantity of the moss peat was used. It should be laid down from three to four inches thick.

And again, horses will often eat clean wheat straw, if they have an opportunity, but there is no fear of them eating the moss peat. Some people say the latter is likely to heat in the stable. This is not the case if it is used properly. Of course it will get hot if it is allowed to accumulate, and the wet is not taken away frequently. My idea is that it does not take anything like the time to keep the stable clean and nice when the moss peat is used, as it does when the horses are bedded down with straw. I will admit the former does not look so nice as the latter.

Referring to long straw again, the shorter it is the longer it lasts, because when it is long, a part of it gets wet and dirty, and it looks bad if left in the stable, therefore it has to be thrown away, when not more than the ends of it are just soiled. This is really a great waste. If gentlemen, and others, who occasionally have their horses go lame, without any apparent cause, would try an experiment, they would find by using moss peat in the stable a number of accidents would be prevented. If they do not use it entirely, I should advise them to spread a little over the bottom of the stable floor, and then put the straw on the top of that. It is nice and soft to the horse's knees, and at the same time gives the animal a good foothold when getting up.

When a gentleman gives from £50 to £250 for a horse, it is worth taking care of. Many stablemen object to the moss peat being put under the horses, on account of its being dusty, but I do not find it gets on the skin of the animals at all, only on the surface of the hair. My horses' coats look quite as well now I use the moss peat as they did when they were bedded down entirely on straw. I am very particular about my animals lying comfortably, I could not sleep if I thought they were not alright. If the owners of horses were to give this a moment's thought, I think they would study them more in their bedding down.

It not only prevents them laming themselves in a number of instances, but it adds so much to their comfort.

When a horse has done a day's work, the more comfortable it can be made, the better. Many horses, after they have scraped away the bedding from under them, dare not lie down, especially old horses, on account of slipping in trying to get up. I have known some middle-aged ones, where they have pulled the bedding about in this way, not to lie down more than once a week. Young horses do not mind much, but I have known old ones myself to stand twelve months right off. One old horse in particular I could name that had not laid down for four years. This animal was over thirty years of age. They have been known not to lie down for five or seven years, except they have fallen down when standing asleep.

A horse, when in good health, does not lie down for more than from two to three and a half hours at the outside, as they get the cramp very often, if they lie in one position for very long. Sometimes they will lie down three, four, and five times in one night, so that it is well to be careful, and see they are comfortable in the stable, and can get proper foothold. It is usually noticed that a horse rests very much better when put in a loose box than when tied up, for the simple reason, the whole of the loose box is well bedded down, and they get proper foothold when they try to get up. When they do this they are not so timid, and rest much better.

There is another mistake which I think many people make. When horses stand in the stable all day, the custom is to bed them down with straw at night, then in the morning after the stable has been cleaned out, the straw is either taken away, or put under the manger, and the horses have to stand all day. Occasionally a young horse will lie down on the bare bricks or floor, but it is very seldom old horses will, as they know the danger of getting up when once they are down.

Now my idea is always to bed a horse down in the daytime, to induce it to lie down, rather than encourage it to stand. I find that the more a horse rests his body in lying down, the longer the legs last. For instance, I have a cob at the present time, an engraving of which forms the frontispiece, which I have driven nearly seven years. Five years he did from twenty to fifty miles in a day, never having a rest except on Sunday. If I took the horse on a journey, whatever stable I put him in I always had him bedded down in the day-time, just the same as he was when at home. Between his feeds he would lie down, and I found he did just the same away from home.

When horses are on a journey, if they can lie down an hour or two in the day, it is a very great help to them. It is the same when a man goes a long walk, he rests much better when he lies down than when he stands. I know it is not the rule for horses to lie down in the middle of the day in a strange stable, because they have never been accustomed to it. A horse will stand in the stable more than double the time without swelling in the legs, if it is accustomed to lying down in the day-time.

The horse I referred to is as perfect in his legs as he was when six months old.   Some readers may think he is a slow jogging horse, and does not go fast.   There is certainly something in that.   Of course, a horse which goes along slowly usually lasts longer on the legs than one which travels at a high speed.   This cob of mine has many times done a mile in less than three minutes trotting, and not a day passed without his doing a mile in less than four minutes, that was doing five and six miles together.   So we cannot call him a slow jogging horse exactly.

Take cart horses for instance.   On a Sunday, when they are not at work, they are bedded down, and if we take a peep in the stable between half-past ten and half-past twelve, supposing . there are ten horses, we shall find from four to eight out of that number lying down, sometimes every one of them   Many gentlemen will not have straw put under the horses in the day time, and I repeat again I cannot help thinking this is a great mistake.   Horses have a great weight upon their legs, and the more they are encouraged to lie down the more fit they are for work.   I have known one or two cart horses which went blind and grew very restless in the stable, and appeared afraid to lie down after they were six years old ; they were worn out after they were from twelve to fourteen years old.   Those which have been in the habit of lying down have been finer horses.

and able to do their work better at twenty years of age than those which have not been used to lying down would at fourteen years old.   When I say horses do not lie down at all I mean after they have become old and blind.   I have known horses live to be twenty-five and thirty-five years old ; one I have in my mind now is thirty-three next foaling.

This book, as I said in the commencement, is not so much for those who keep a number of horses as for individuals who go in for the ones and twos.   In the cold frosty weather when a horse is standing in the stable doing nothing, and getting very little exercise, it becomes very cold and the hair stands on end, even when it has a cloth on it.   When a horse's legs are cold the ears are usually the same, and when the latter are cold it is an indication that the horse is cold all over.   I have noticed that when a horse is cold in the stable, after it has laid down for a couple of hours, it gets up quite warm.   This is another reason why horses should be encouraged to lie down more.

I may explain why this is.   Horses have nothing upon their legs, beside bone, sinew, and the skin, so the blood circulating down the legs becomes cooled and flows back through the heart, and gradually chills the whole body If the horse lies down the circulation is easier and more rapid.

CHAPTER V.

CLEANING HORSES.

How Horses are sometimes cleaned—How they should be cleaned —Attention in wet weather.

MANY people who go in for keeping a horse, or horses, have different ideas as to how the animal's coat should be cleaned, and kept in good trim.   It is a well-known fact that numbers of grooms spend close on an hour in grooming a horse.   One thing struck me in my younger days, and I often wondered at it.   The farmers' horses frequently had better coats during the Summer than gentlemen's horses which had six times the amount of labour bestowed upon them.   The attendant who saw to the farmers' horses would have from six to ten horses all cleaned and ready for work by six o'clock in the morning, and yet most of the animals' coats would look like a piece of silk.

I have noticed many people, when cleaning horses, using a curry comb, and giving the coat a good rake all over, when doing this they often work the hair the wrong way. That is to say, instead of currying the way the hair lies, they scratch the opposite way.   If these people are asked why they do it, they tell you it is because it gets the dust out quicker." When a horse comes in all covered with perspiration this would answer, as naturally the dust works out of the skin on to the surface of the hair, and the curry comb takes it off quicker, but when the horse is cool, and has no harness marks upon him with the dust, it is not well to use a curry comb, at least that has been my experience.

The more the horse is curry-combed, the more scurf is produced on the surface of the skin, so that excessive curry-combing means an excessive amount of scurf, especially in some animals which seem to have a tendency in that direction. If instead of using a curry comb, a dandy brush, or better still, a body brush is used, it will be found much better, especially if the horse has been working, and there is a lot of surplus dust on the skin. Some people use nothing but a wisp of hay or damp straw. The man who attends to farmers' horses will not wear out as many brushes and curry combs on ten animals as a gentleman's groom will on one, because the farmer's man uses a damp wisp of hay, or straw, and the curry comb but very little.

I have tried both methods myself, and find a good hard rub with a wisp of hay, or straw, damped a little, cleans the horse's coat far better, and leaves a better gloss on it than anything else can do. A body brush should be used more than a dandy brush. I think most of my readers will know what I mean by a body brush. It is a flat brush, with a strap across the back to slide the hand in. It is soft, and does not work the skin so much to irritate it, but thoroughly cleanses it from the surface dust.

I am quite aware that some horses have a scurfy skin and require a great deal more cleaning than others, but even in such cases as these it would be better to use a curry comb much less than many people do, as it only irritates the skin of the animal.   When a horse has a great deal of scurf it should have a mixture given it, made up of about a pint of stewed linseed, a little bran and chaff, with two table spoonfuls of common sulphur.   If this is given the animal twice or three times a week it will prove a great help to it.   Of course when a horse has this mixture given to him he must not be allowed to stand about and get cold, as sulphur opens the pores of the skin; if this is seen to it will be alright.   In the Spring nothing will cool the horse's blood so much as nettle tea.   This is easily made by pouring a gallon of boiling water on a good double handful of nettles.   This is not only very cooling to the blood, but puts a nice gloss on the animal's coat.   Some horses will object to drink this, but when they get accustomed to it they like it very much.   If a horse is in very low condition it should have a tonic, especially when it is changing its coat.

Horses are sometimes allowed to run down in condition when a little medicine or tonic would put them alright, and bring them into splendid condition.   Of course horses living in the country that go out to grass do not require anything of this, but those which cannot get such a luxury sometimes require a little pulling together. There are many different opinions as to whether horses should be washed when they come in all of a lather on a hot summer's day.   My experience has been it is a good thing to wash them directly they come in, while they are hot, with just chilled water, and scrape them down with an iron hoop, so that all the water is scraped out of the coat afterwards.   A little brass scraper is sold by saddlers, which would be better than the hoop for the purpose.   This scraper can also be used for all kinds of horses, when they come in wet on a rainy day, to get the water out of their coats, then they dry very quickly, as if left the evaporation of the water takes very much longer, and some horses occasionally take cold if left in this way.   Then if they are well rinsed, and rubbed over with a sponge and wash leather, they dry much quicker than if they had never been wetted at all.   As a rule, the perspiration from a horse is of a greasy nature, the water washes all that out, they dry much quicker, and are more comfortable afterwards.

Bandages should next be put on the legs at once, and if they are accustomed to have a cloth put on them whilst standing, one should be thrown over them directly they are dry.   If the weather is at all chilly when a horse is washed it should always have a rug put on at once, at the same time taking care to put a good piece of straw on the horse's back under the cloth.   This prevents the animal taking a chill, as the straw absorbs the evaporation.   In some instances the rug gets wet and is not dried when it is taken off.   If this is put upon the horse again damp, it is liable to give it a chill.

Some horses have such long coats during the winter,

it is almost an impossibility to keep them nice, as when they get so hot and perspire the hair becomes rough and wrinkled.   I deal more fully with this matter in the chapter on clipping horses.   Sometimes when a horse is neglected in the way of cleaning, it causes sore shoulders through dust and perspiration being left unheeded on the animal. This is often the case with many costermongers' horses and others, which have no care taken of them.   Then again, there is another thing which often happens when people take no notice of the horse's collar.   That is the dust becomes matted very hard inside the collar, making it very rough, which often causes the animal to have sore shoulders.   This is especially seen in the case of cart or other draught horses, also in those used by butchers, bakers, and costermongers, who are oftentimes so busy that they take little or no heed whatever until they find the horse's shoulder is sore.   This is not only cruel, but expensive to the owner, as the animals are often laid by for a week, or perhaps a month, whereas if a little forethought were used all this might be avoided.

D

CHAPTER VI

CLIPPING AND SINGEING.

Reasons why horses should be clipped—How, when, and where they should be clipped—Clipping not always to be followed by singeing.

THERE are various opinions expressed on clipping horses. Some people say it is not natural to take off the horse's coat. This may be true there are many things not natural, but still beneficial. This is my idea with regard to clipping horses. I have tried the experiment in many ways, both with those which run, and cart and waggon or draught horses. As a rule I find animals which are clipped do their work with much more ease than those which have their long coat on. Of course many thorough-bred horses have an exceedingly fine coat, are well groomed, and always rugged up in the stable.

In such cases they may do without the clipping, but I am speaking now of the general run of horses. Those which have to work and sometimes run twenty miles right off with a long coat on often become very faint and tired, especially if the roads are heavy. When the animals get in this state the long hair is so wet that it will take from three to six hours to dry after it gets in the stable

D2

Now I cannot see how any one can say this is good for a horse. Take human beings for example, when we have to walk or run do we not like to be as lightly clad as possible, then when we have finished our journey we put on a big coat. If we run in a heavy coat we soon perspire freely, so it is with horses. I maintain, those which run do better if they are clipped: and these, after they come in, should have a rug thrown over them. If they are allowed to stand while out of doors they should always have a rug thrown over their loins.

I know I differ from many people as regards clipping their legs. I do not believe in doing them below the knee joints, unless the horse can have good attention. For instance, when it comes in wet and dirty and is washed or rubbed partly dry, then bandaged up, I do not object to the legs being clipped. In the ordinary way, however, when they cannot have this care and attention, the legs should not be clipped below the knees. I find they are so much more subject to mud fever when this is done than when the hair is left on them. Veterinary surgeons tell us mud fever is brought on through a chill in the blood. When horses get the hair clipped close to the legs in cold weather they feel it very much. No doubt this is owing in a great measure to the animal's legs being so void of flesh, and nothing but sinews, skin, and bone. The blood circulates down the legs, therefore, if the latter are cold, the blood is carried back through the body chilled. Gentlemen who do a great deal of hunting, especially of late years, seldom have their horses' legs clipped.

Horses which are restless in the stable, and pull the bedding-down back with their fore-feet before lying down, are far more liable to scratch their knees, or rub them when they are lying down. I recommend that horses which run should be clipped all over except the legs, but tradesmen's animals, such as butchers, bakers, and grocers use should be clipped trace-high, because of their having to stand about, as there is not always time to put a rug on the animals. It is very bad for them to be standing after they are clipped, especially when it is raining, or is a very cold day. If the hair is left on the back and legs the animals do not feel the cold. Many people have their draught horses clipped just in the same way, as they do so much better and get through their work easier. Of course this is only done when draught horses have on a heavy coat ; some of the hair is much finer than on others.

Various opinions are expressed as to when a horse ought to be clipped. This depends a great deal on when the animal sheds its coat. Some horses shed their coat much earlier than usual in the Autumn. Many horses get quite a winter coat by the first week in October. When the summer coat is shed, and the other is fairly long in the Autumn, and the horse is working very hard, it becomes very faint. This is the time to clip the hair off. I always like to give the animals a little tonic while changing their summer coat for the winter, as at that time they go down in condition more than at any other period.

When they are changing the winter coat for the summer it does not matter so much, they do that with comparative ease, but it is just as well even then to give them a little tonic. Those who understand horses usually give them something when they are shedding their coat, whether the owner gives instructions for them to have any or not, and in many cases these men know what the animals require better than the owner. Some horses require clipping twice during the Winter—I often have mine clipped four times during that time, but, as a rule, twice is sufficient.

It is very seldom a horse's coat grows after the middle of December, only just a few rough hairs underneath the stomach and round their quarters. Those who wish to keep a horse nice should just singe these off. Years ago it was thought a horse which was clipped should always be singed, that is to say, the end of each hair ought to be singed, and unless this was done it was considered the animal would take cold, but this is only a delusion. I have never known a horse to take cold after it has been clipped, if it was managed and clipped properly.

Some people do not like clipping their horses because it changes the colour of their coat. For instance, there is very little difference between the colour of a bay and a black horse when they are clipped. When they change their coat in the Spring it makes the horses look rather stubby, but that very soon wears off, especially when they have a little tonic to help them. This not only enables them to shed their coat quicker, but makes the new hair shine like a piece of silk in a very short time. Care should be taken before a horse is clipped not to rug it up for a time, or it should only have a thin rug thrown over it, and after it is clipped it must have a very warm covering on in the stable.

CHAPTER VII.

HORSES' FEET.

The foot—What causes lameness in the stable—Observation and explanation—Corns and mud fever—Goose grease and water for horses' feet—Thrush and its treatment—Cracked heels.

Perhaps there is no part of a horse neglected so much as the feet, and as it is so very important that these should be well attended to, I will give a chapter on this subject for the benefit of those who do not understand horses and very seldom give the feet a moment's thought. Many hundreds of horses are ruined every year simply for want of a little care being taken of their feet. I will endeavour to make myself as plain as possible in this respect. Now it must be remembered a horse's foot is wonderfully constructed. Nature has provided it with a thick horn for a covering round the outside, the middle part is covered with a substance more elastic, which is called the frog, and the foot itself is enclosed between the two, while the sole of the foot is different from these two, being formed of a soft horny substance.

A great many people think that a horse's foot is simply a foot, and only made for work. They not only think so but act accordingly. They merely feed and work the animal, and when its shoes are worn out send it to the farrier to have some more put on, not thinking whether the foot is in a healthy state or not.   I will not say they do not care whether this is the case, because they really do not give it a thought.   As long as the horse does not go lame and does its work alright, not more than five out of every twenty persons who keep horses ever examine the hoof or foot.   Of course, as soon as the animal begins to limp and show symptoms of lameness, or tenderness in stepping, then the owner at once examines the horse in the legs, shoulders and feet, to see what is wrong.   I always say, " prevention is better than cure," though it is not always thought so. Horses costing from £40 to £150 go wrong for want of a little care and attention to their feet.   I do not say one here and there is spoiled, but hundreds every year, besides many others of smaller value, viz., those belonging to butchers, greengrocers, costermongers, &c.   If only these little things were seen in time how many valuable animals might be saved.

As soon as a horse begins to go wrong in the feet, or have what is called fever in the feet, the veterinary surgeon advises the owner to sell it as soon as possible. This may be all very well in some cases, but in others it is very unwise.   I saw one splendid carriage horse sold for £9, and another hunter sold for £15, only during the last eighteen months.   With care both of these animals have got round, and are now splendid goers and show no signs of any tenderness in their feet either when running or at work. Neither of the owners would part with them now under a long figure.   I happen to know both the gentlemen who bought the horses, and saw the animals at the time they were sold, though they were both disposed of in different parts of the country. I asked the owners why they bought them in the condition they were in, as they walked like a cat on hot bricks. The answer I got from them both was, " Oh, the horses will be alright in a very short time, they only want a little care and attention." Of course, it must be borne in mind both these gentlemen understood horses thoroughly, and had bought many such animals, then after they had come round they fetched very long figures. I just mention this circumstance as an illustration to show how people are mis-led by horses becoming lame in their feet. In some cases the horn which goes round the top of the foot becomes very brittle and hard, which of course binds the foot, and I have every reason to believe that corns often appear on the horse's foot simply through this. Then when the horses are running on hard roads, the jarring causes these to form quicker, and makes them grow faster, as the horn presses on the foot so much and the animal has to work hard at the same time. If a person wears a very hard pair of boots which press tightly to the feet, it brings corns, and so it is with the horse. That is usually what a blacksmith looks for when an animal goes lame in the fore feet. He cuts the sole or bottom of the foot away to find the corns, and digs in close to the hoof, making the foot more tender than before, and after the sole has been cut away so closely and then a hot shoe put on, this causes inflammation to set in at once and makes the horse a great deal worse than it was before. When the horse goes to be shod these corns have to be cut out every month so that there is no pressure from the shoe.

A corn is a thing which grows very fast, and in three weeks or a month's time, very often before the shoe is worn out, the corn is so far developed that it often hurts the horse before the shoe is removed. The owner of a horse which has corns should always see that they are cut well back. Those who have animals which have no corns should endeavour to keep them free from them. This can be done in most cases by keeping the hoof in good order. When the horn is neglected or becomes hard, the foot gets hot and dry, then if it is not seen to for a few days it brings on fever, and the horn or hoof becomes very much contracted, so much so that it will often show ridges on the surface of the hoof. This means unsoundness in the foot at once. If a horse's foot is not taken care of, the horn gets so dry and brittle that it breaks all to pieces, and when the farrier nails the shoe on, it often splinters like putting a nail into brittle wood. When however the foot is seen to properly, it is very seldom this occurs, as it keeps tough and elastic. A horse's hoof should always be kept fairly clean, and it does it good to soak it in water, especially in hot weather when the horse is running on hard roads.

Horses used by gentlemen usually have greater care bestowed upon their hoofs than ordinary horses which are only used for business purposes, as they generally have a little lamp black and oil rubbed over them. Goose grease is a fine thing, and a little goes a long way. The hoofs should be washed perfectly clean, then the grease is rubbed in with a small brush. It answers two purposes. First, the feet always look nice; secondly, the horn is preserved and kept in splendid condition, so much so that it rarely gets broken, as it is tough and pliable. The mixture has the same effect as moliscorum on hard leather. I have met with several grooms who use lamp black mixed with fat skimmed from a dead horse which has been boiled. I have never used this myself, but know several who do use it, and it appears to answer very well. I have seen the hoofs of horses which this has been used upon and they have been in splendid condition. Now the frog of the foot is found to be of quite a different material from the outer horn. Years ago farriers always used to cut the frog down very close, also the crust or sole of the foot. This is very wrong, as they should be cut down a very little, just trimmed so as they are not shaggy.

Farmers' horses which work on the land seldom have anything wrong with their feet, first, because the ground is moist, which keeps the foot cool, and secondly, when they go to water they often soak their feet in a pond, which helps them very much. Waggon horses which are on the road do not go lame in their feet so quickly as horses which run, because the jar of the feet in running makes all the difference. Of course cart horses which are often working on the hard road go lame much quicker than farmers' horses which work on the land, in fact it is very rarely the latter ever have anything the matter with the feet. If I had heavy horses working upon the hard roads during the hot weather I should let their feet stand in water for at least five minutes before they went out in the morning, and again when they came home at night.   It would be well, where a person has a number of horses, to have a place made where they can go and stand in for a time when the weather is hot: it would pay the owner 100 per cent. for the trouble.   Such a place could easily be made with cement, nice and shallow, so that the horses could walk into it easily.   If the water is laid on, have a tap for supplying it, and a little plug at the bottom with a drain to empty it.   In country places there is usually a pond where horses can stand in.   Water has a great effect, not only upon the feet, but on the legs also.

I have known a great many horses bought in London where they have been running on stones, and their legs and feet have been so bad the animals could scarcely stand upon them.   When they are like that, farmers often buy them at the sales in London, take them down in the country, turn them out in some damp meadow, and let them stand in a pond from one to four hours a day.   In many cases the legs and feet get quite cured, and as perfect as they were before.   Some people, instead of sending them on to the sales, kill the animals off at once.   Horses which run should always have their feet washed before they go out of the stable.   It is always better to serve the heavy horses in this way as well, if time will permit, but it is very seldom these get washed at all, as it is not usual.   This is only done on a few particular farms.   Soft soap, nice clean water, and a good stiff brush, are the finest things to wash horses' feet with, the soap is very cleansing.

Another thing which people have to guard against is what is commonly called "the thrush." This is a running or discharge of a thick matter from the centre of the foot out of the frog. It cannot always be said what brings it on. Sometimes overfeeding with corn and too little exercise will do it. The discharge is very offensive, and if not attended to spreads over the foot, the centre of the frog always appears as though it were rotten, and it naturally affects the foot very much if it is allowed to grow. Now I always look upon the thrush in a horse's foot as an outlet of Nature, and if it is stopped all at once it will break out somewhere else, or the animal will be ill. I have known cases where horses have become quite blind through the thrush being stopped quickly. It is the same as a sore upon a person's finger or body, there is a certain amount of matter comes from the wound, and so it is with the thrush in horses' feet. A person can always tell when it is coming on, as the foot begins to smell. Both the foot and centre of the frog should be cleaned out once a day with a hook, and sometimes it will be noticed there is moisture on the hook which smells badly : that is the thrush just commencing. Now this should not be stopped at once by any means. The dressing should be of that nature so that it does not close up the wound, but has the power to draw the moisture out, and prevent its spreading over the frog when it does come out. When a horse begins to show symptoms of thrush it is well to use simple remedies. I will give one which I have found very useful. Put two table-spoonfuls of chloride of lime with a pint-and-half of water, and if it smells very badly, and the moisture begins to run, use a pint of water instead of a pint-and-half with the chloride of lime in.

Some people prefer a little salt put in, no doubt that is a good thing, but I never use it myself. While the horse is standing in the stable, soak a little piece of tow in this lime and water and put it in the frog with a blunt knife, or a piece of wood cut in the shape of a knife. Before taking the horse out of the stable remove the piece of tow, and wash the wound with a little brush dipped in the water and lime, then get a small piece of wood, just cut so that it will go nicely into the frog, and dip it in Stockholm tar and place it in the wound. The tar draws it and prevents any smell coming from the horses' feet. Some people use nothing for the thrush except the Stockholm tar, while others use bluestone, but this usually dries it up too quickly.

Many horses have the misfortune to get cracked heels. This, however, seldom occurs when they are washed down and well attended to. This is usually brought on through neglect. Very often when a horse has long hair on its legs the heels are not noticed, I have seen them with a crack half an inch deep, before the owner or attendant knew there was anything the matter with the heel at all. This is caused by the legs not being rubbed dry, the water naturally runs down to the bottom of the leg and round the heel. Should there be a little dirt there, as there often is in neglected horses, the cold water mixes with it, and that, with the friction caused by the animal moving its foot, causes the skin to get hard and brittle, and it very often cracks. When this is the case, if not properly attended to, the crack soon becomes very deep, so much so that the animal is often laid up with it. Almost in every case where the heel cracks there is a discharge from it, and if that is neglected the hair gets matted every time the heel becomes wet. This of course makes the heel a great deal worse, and the smell grows very offensive if it is neglected for long. When it breaks out in the heels upon the hind legs it sometimes turns into quite a humoury mass.

I have known this complaint to break out in a cart horse and grow into a greasy heel. That is to say, the leg has swollen very much and the heel has become full of humour. When a horse once gets a greasy heel very badly, it is seldom cured. I have seen them swell to three times the size they should be. If a horse has no humour in it, greasy heels will never follow a cracked heel, although they may be very sore and troublesome. Some horses are naturally full of humour in the legs. I have known mares which have a greasy heel—which some people call a swollen leg—throw some colts with legs equally as bad. In such cases as these it is in the blood, and is hereditary. This very seldom happens with nag mares, it is more often with cart mares. The way to prevent cracked heels in horses is to dry the legs well after they have been wetted, and if they show tenderness in the feet a little fat out of a fowl is one of the finest things which can be used. It is of a healing nature, and makes the skin soft and pliable.

When a horse has a deep crack in the heel, which begins to run with matter, it should have very little corn,

E

but should be fed principally on stewed linseed, bran, and hay. Stewed linseed and bran mixed with two tablespoonfuls of sulphur cools the blood, and if the fat from a fowl is rubbed into the wound it soon draws the matter out and heals it up at the same time. Some people put alum on a cracked heel, which is very wrong, because it has a tendency to dry and harden, and although it dries the place for a time it makes it hard and brittle and very often it will crack again directly. A little dry fullers' earth put into the wound is an excellent thing, as it is of such a healing nature, and dries it at the same time. If a little fat from a fowl is used it makes the skin pliable. This will generally cure a cracked heel in a horse, no matter how bad it is, and often without the animal lying up at all.

Whenever a horse has cracked heels the legs should never be washed more than is really necessary. Cracks are not like ordinary sores, because there is a kind of friction, and every time the foot moves it opens the wound, which at once begins to bleed or discharge matter. Sometimes too much corn and too little work will make the horse humoury, and this is likely to cause a break out in the heel. The remedies I have mentioned are simple but very safe ones. A greasy heel, properly speaking, is not what is called a cracked heel, though the former may arise from the cracked heel. I have said the latter contains a great deal of humour, and when a horse has this complaint it should be kept very low and have but little corn and only do light work. If the leg and heel are well rubbed with the fat of fowls, and a little fullers' earth is put in, it will draw a great deal of the humour out of it, and in many instances will quite cure it. It is far better to prevent this from occurring by exercising a little care.

CHAPTER VIII.

BANDAGING AND MANAGING HORSES IN THE STABLE.

Bandaging when on a journey—Preparation for binding and application—General directions for cleaning and drying—Ventilation.

Many people believe in bandaging horses in the stable, while others have a great objection to it. My experience teaches me that after a horse has done a hard day's work there is nothing like it. This refers more particularly to horses which run; we do not think of bandaging a cart horse unless it has a bad leg. When a horse works hard, it shows a little puffiness in the leg, which some people call wind galls. I often find the sinews of the leg swell through hard pressure and the jarring on the hard roads. They often become extended, especially in the fore-leg, which causes the horse to stand weak upon the legs. This is what is commonly called "over at knee." Veterinary surgeons may differ from me in this respect, but, nevertheless, I believe it to be true. I find if a horse is bandaged when it comes in from a hard day's work the legs do not swell, but go down; neither do they look so puffy the next morning. It is the same with a man who has been working hard all day with his hands, the sinews of the wrist are swollen and sore the next morning, and when he goes to work again they hurt him so much he has to bandage them. So it is with the horse when it has been working very hard; the next morning there will often be noticed a puffiness about the legs. Now if the latter were washed down and well soaked with water, and at the same time some flannel bandages were put on carefully, it would prove of great benefit to the legs, in fact, it is only those who have tried these bandages who really know the value of them.

A gentleman who has a splendid hunter, which can run and jump well, will take care to have it bandaged directly it comes in from a journey. When I am driving on a long journey I always take bandages with me, and if the horse is going to stand in the stable more than an hour I have them put on his legs. To test the value of them a person should bandage one fore-leg of his horse and one hind-leg, leaving the other two just as they are. If the horse is in frequent work the difference will soon be noticed, those which are not bandaged will become very puffy and look altogether different from the other two. I have tried the experiment myself, and if any of my readers do the same I feel certain they will never allow their horses to stand in the stable after they have been hard at work without bandaging the legs. The legs of a horse at twelve or fifteen years old are as clean and as nice as a young colt four years old if they are kept bandaged and are looked after as they ought to be.

I have seen horses from five to seven years old puffy in the legs and moving very stiffly when going out in the morning, looking as though they were worn out. After careful treatment in bandaging their legs and soaking their feet in water, in twelve months' time their legs have been perfectly right again. When a horse shows weakness or puffiness about the legs it is well to use a little embrocation and rub it well in. If the attendant has none by him, a little vinegar and water helps the horse very much if it is rubbed well in the leg and then bandaged at once. The legs always look nice and smooth if treated in this way. In the Summer I believe in using plenty of water for the horses legs, in fact, if they are really weak, I often use wet bandages. In cold weather, if the legs are very dirty, wash them down and dry them as much as possible with a sponge and wash leather, then rub them down well with a wisp of hay or straw and put dry bandages on at once. A horse's legs become dry quicker while he is warm, and they should never be washed in the Winter when he comes in, either day or night, unless he is bandaged at once, if so, it is liable to cause a chill, which at times brings on mud fever. I do not say it will always do this, but it does very often.

As the legs of a horse are very little but skin, sinew and bone, and the blood circulates through the legs and feet, they soon become cold if they are washed down with cold water in the winter months, and the blood is carried back to the heart slightly chilled. This causes "the shivers" in a horse. No matter how dirty a horse is it is not always wise to wash it. It should just have the dirt rubbed off with an old brush or wisp of hay, then the legs soon become dry and the animal is warm and comfortable. Of course it makes a difference where there is a groom kept specially to look after the horses, so that when they come in they can be washed down and made dry and comfortable. I am referring here more particularly to those who are in the habit of bringing the horses in the stable on a cold winter night and just washing the legs down, never thinking or troubling about drying them after. Many horses are ruined in this way, and those which are not usually get mud fever. I have seen horses with lumps on their body the size of a small nut, which have been caused simply through the animals being wet and getting a sudden chill. A difficulty may crop up here in some people's minds as to what should be done when a horse comes into the stable on a very wet night in the Autumn or Winter. In this case the animal should be scraped down with what is called a scraper, which can be obtained at the saddler's. Those who do not possess one can make a small iron hoop off a cask answer the purpose. The water should be all scraped off the horse clean, and the animal well rubbed down with a wash leather. Some people may not have a wash leather at hand, but some nice dry pieces of straw will answer the same purpose. If the horse has been accustomed to being clothed up, before the cloth is put on, little pieces of straw should be laid right along the back, from the tail to the neck, then the cloth should be put on the top of that. When this is done the steam from the horse's body goes into the straw, which keeps the back very warm, and naturally causes the animal to dry much quicker. If the straw is not put on first, but only the cloth, the steam from the horse's back goes into the cloth, and while the body is drying, the former is getting damp, which often gives the animal cold when the rug is put on again after being taken off damp.

If only a little care is taken in this respect it saves a deal of trouble in the long run. Where a man thoroughly understands his work it is very seldom a horse takes cold while standing in the stable, as the animals are looked after as they ought to be, and clothed up with straw on the back in the way I have described. It is very seldom waggon and cart horses are clothed up when in the stable, at the same time they get wet just the same as nag horses. They should be well scraped down, and rubbed with some straw, and, as a rule, they very soon dry as they stand. Many heavy horses catch cold through coming into the stable cold and wet, and never being attended to properly. It must be remembered a horse which has a long coat, and is not scraped down, holds a great deal more water than one which is kept clipped short. The horsekeeper should always bear this in mind, and see that the water is well scraped off the horse. If the animal's ears are cold they

should never be left in that state, because it is an indication the horse is cold all over. They should be rubbed
through the hand until they become quite dry. If they
are warmed in this way it usually circulates the warm
blood right through the body. There is nothing a horse
enjoys more than having its ears rubbed when it gets
accustomed to it. When the owner, or attendant, goes
into the stable to see whether the animal is comfortable
or not, he should first go and feel its ears. If they are
cold and wet it is a sign the horse is not left comfortable.
No matter how hungry the animal is, if the attendant
commences to rub its ears through his hand, so as to
bring circulation into them, the horse will hold its head
right down to the ground in order that the ears can be
reached nicely. The horns of a cow should be felt to
see whether the animal is cold or not, and the ears of
a horse.

I mention rubbing heavy horses, because they are as
capable of taking cold as nag horses, though not quite
so readily. Some people may argue that when horses
are lying out in the field they are never rubbed down
after it has been raining, yet there is not one in a hundred
takes cold, even if they are lying out in the winter months,
and are not brought in at all. That is quite true. It is
natural for a horse to lie out of doors. But suppose a
horse was working and was then turned out in the field, it
would have quite a different effect on the animal if it were
to rain hard, that is if it were not attended to. When
horses are doing no work, and are lying out in the wet,

no matter whether it is in the Winter or Summer, they very rarely take cold. This is accounted for by the fact that the blood keeps about one heat. But when horses are working their blood gets much hotter, and if they are not taken care of when they come in wet, their blood gets chilled. I do not believe in pampering a horse up, shutting him up in a warm stable and giving him no ventilation. Horses want plenty of air, and if a little care is bestowed upon them they will well repay the owner for the trouble.

All stables should be well ventilated if made and kept warm, to allow the hot air and fumes from the urine to escape. I cannot say just how the ventilators should be arranged, because designs and situations differ. The ventilator should always be put in at the highest point possible, so that no draught is allowed to come on the horses. More horses catch cold owing to the badly ventilated stables they are kept in than from any other cause.

CHAPTER IX.

COLDS, DISEASES, AND SHOEING.

Colds—Care in Shoeing—Watch the Farrier—Shape of ordinary
Shoe—Patent Shoes.

HORSES, like other animals, are not exempt from
diseases and accidents, but it is not my intention to
treat upon the various complaints they are subject to
because when a horse is really ill it is far better to call in
a veterinary surgeon at once. I do not give advice I
would not take myself. If my own horses are ill, although
I understand a little about diseases, I make it a rule,
because I think it wise, to call in a veterinary surgeon at
once, as delay is often dangerous. Of course, I do not
mean to say when a horse takes a little cold it is necessary
to send for a veterinary. If it only happens to be a cold I
never use anything but stewed linseed, worked up in a bran
mash, and let the animal have it as hot as it can take it.
If the horse has a sore throat the steam from the mash
does it a great deal of good, and at the same time the
throat should be well rubbed with a little strong oils
between the two jaw bones and round the gullet. A
person who does not understand horses should always
use simple remedies, if they do not get proper advice

at once. For instance a horse sometimes ricks itself in getting up in the stable or sprains its leg or foot while at work. When this occurs, do not work the animals, but give them a little exercise each day. Let them have but very little corn, as their blood wants keeping cool when they meet with an accident in the way I have described. Sometimes a horse will get mud fever. This usually comes on through a sudden chilling of the blood, and little spots rise up in the skin. If the horse is not kept cool for a week or two, these little spots or lumps usually break out into sores. There is no danger with these, only the animal should have but very little corn and no beans, peas or maize : at such times as these it should have nothing but bran and oats. Corn, beans, peas, and maize are all more or less heating to the body. If a horse has poll evil, fistula, or any complaint of that kind, it is far better to call in a veterinary surgeon at once. Some horses are very subject to sore shoulders, and it is well for the attendant to always notice the shoulders when a horse is working hard. It is possible for one to be working all its life and never have a sore shoulder, at the same time many horses are subject to this, especially when wearing a new collar, or when the animal has not been working for a week or two. For instance, in bad weather, if a horse has not been doing anything for say two or three weeks, when it is put in harness it is rather fresh, and the shoulders are a little tender, sometimes they become very sore ; should this escape the eye of the attendant, the next time the animal is worked the skin comes off and leaves a raw place. When a horse has been standing in the stable for a fortnight or three weeks, no matter what sort of animal it is, the safest way is to rub its shoulders with salt water, which hardens the skin very much. Where the hair and skin have come off nothing will do it more good than washing the place with a little water and fullers' earth. Wipe the wound dry before putting this on. A small piece of goose grease put on the place is a very good remedy, then if a little fullers' earth is sprinkled over the top of that, it causes the skin to harden almost at once. This treatment is good for the cure of almost any sores which may come on horses.

Many of the animals are lamed through careless shoeing. Farriers often get so accustomed to nailing on the shoes that they occasionally prick the animals. Particular care should be bestowed in shoeing horses which run, even more than on farmers' horses and those which draw heavy loads and only walk. When a horse is being shod it is well not to have too much of the sole of the foot cut away. Some blacksmiths pare it off almost close, but this is wrong. Owners of horses should always take particular care of their animals in this respect, and see that the farrier or blacksmith does not pare its foot down too close. I have seen them cut down so that the blood veins could be seen, and in some instances the feet have bled. Now this is very wrong, because the shoe only covers part of the foot, and although it keeps the latter from the ordinary road or ground, when the animal comes to any rough or sharp stones it will often become lame through stepping upon one, as well as getting fever in the feet from a bruise by so doing, and in some cases where the feet are tender they go down. As regards what sort of shoe should be used for certain horses that is difficult to tell, as most people should know what work they are going to put the animal to, and be able to judge what kind of shoe is required according to the amount of work to be done. For instance, a horse which runs a good deal should not have the shoes too short in the front, if so, it causes weakness in the fore leg.

With a horse which goes hunting, it is a different matter : the fore shoes must be rather short, if not the animal pulls them off with its hind feet when it goes to jump, but most gentlemen who use horses for this purpose are well aware of this fact, and have their animals seen to properly in this respect. I might mention, when it is sharp frosty weather it is a wise plan to have the horses' shoes made so that screws can be put in by the owners themselves, as it saves a great deal of the animal's time. Sometimes when it is very slippery I have known horses have to wait at the farrier's from two to six hours : this is a great waste of time. Then again, it is very convenient to be able to put the screws in oneself, as they are no trouble at all ; they can be put in with a little screw hammer or pair of pincers, and taken out when the horse comes home. No matter whether they are cart or nag horses they can be served just the same   This treatment is much safer than the ordinary roughing, whether it consists of frost nails or the shoes turned up. If the latter, they are more dangerous than the frost nails, as when in the stable the animal will often cut itself by putting one foot on the other. Horses will often lame themselves in this way. It is much safer and cheaper in the long run to have the shoes made so that the owner can put the screws in himself. Some people may say, "Suppose the frost does not last long ?" Well, that is so much the better. It is not very expensive to have the shoes made so that the little screws can be put in. If there come two or three frosty mornings it pays, but when it comes a long frosty winter these shoes are a great boon. When a horse keeps having his shoes taken off, to be turned up or roughed, it soon weakens the foot and breaks the hoof about a great deal. In fact a horse with a tender foot often gets lamed and for the time being is unfitted for work in consequence. Any farrier can get these screws, they can be had in all sizes, both for nag and cart horses. The shoes can be made so that the screws will fit in properly, and there is no difficulty with them whatever. In the long run it is much safer and more economical to use them.

Some people prefer shoes turned up at the heel, saying it gives the horse extra power going up and down hill. This may be alright for draught horses that do no trotting, but these high heels jar the feet very much when they trot.

F

Many kinds of patent shoes have been brought out during the last fifteen years—some very good ones. One lately introduced by Mr. F. J. L. Clark, consisting of a steel shoe with pieces of india rubber let in the centre of the rim which covers the sole, which he calls the non-slipping horse-shoe, and I think it will be a great boon to horses : as in addition to its preventing them slipping it will save a great deal of the jarring which is caused by running over hard roads. I intend giving these a trial, and my experience of them shall be given later on in " Farm, Field and Fireside," after this book is published.

CHAPTER X.

BREEDING.

The increase in number of horse-breeders—Why farmers should breed horses—Mares rejected for work, but valuable for breeding—Treatment of breeding mares—Selection and pairing—Horse-dealers and their tricks.

ENGLISH farmers breed more horses now than they used to years ago. When railways were first made people said that horses would become so cheap that it would not pay to breed them as there would be so few required. Those who prophesied this, however, made a great mistake : I am told there is more than double the number of horses in England as compared with the time when the first railway was made. I cannot, of course, vouch for the truth of this statement, as I do not remember how things were going on at that time, but I know there has been a large increase in the number of horses in this country during the last thirty years. Many of them come from Ireland and Russia.

I believe in some parts of the country there are three colts bred at the present time to every one that was reared years ago. At that time many farmers would not think of breeding, no matter what mares they had by them, but it is not so now. Those farmers who have a good team of horses can make more money out of them than any other class of people, because they begin to use their young horses at an early age, so that they more than earn their own living. It is a different matter with those who have only a limited space of ground at their command and have no use for young horses. Then again, no animal has been more improved upon during the last ten years than the horse. I put this down to a great extent to the various horse shows which have been held in different parts of the country for the last few years; and in another ten years time we shall have better bred horses than we have at present.

There is a great deal of land lying waste in England, even now, which might easily be used for breeding horses, in fact, a good deal of it has been used for this purpose. Many farmers and gentlemen who understand this kind of business thoroughly go to the auction sales in London and buy splendid mares very cheap, that is—mares which have been worked on the stones very much, or have met with a slight accident. Such animals as these can be bought up at very low prices; I have known them to be got at from £5 to £25 each—really good mares, which have bred from four to nine colts after they were what the London people call worn out mares. When they are put to work in London while young they usually go wrong in the legs or feet.

Such mares as I have described often come in very useful to the farmers, as they usually work them a bit on land, and they do splendidly for breeding from at the same time. I might, however, mention one thing : after a mare has been fed on a great deal of corn, and has been worked hard, she will often not breed the first year unless she is turned out to grass, and has an entire rest; but when she is kept quiet there is little difficulty in this respect. One advantage of horse-breeding is the mares can be worked very well up to the last three weeks or a month before foaling if carefully used, and no heavy loads are put behind them. Care should also be taken not to back them when in a cart, as it sometimes causes them to slip their foals.

If a breeder has a lot of rough land the mares can go out all the winter to eat up the grass; no matter whether they are thorough-bred mares or not they will stand it equally as well as the half-bred and cart mares. It is best to turn them out in the day-time, and bring them in a straw yard so that they have got an open shed to go under if they choose. Many breeders let them lie out altogether and never bring them in at all unless the ground is covered with snow. I think that is a little unreasonable, though, at the same time, they often seem to do well if they have a little bit of rough hay. I like to feel the horses are made comfortable, and it is anything but pleasant when a horse has to be out all night in an open field with the wind blowing and the rain pouring down fast. If they are treated in the way I have described not one in twenty will have a cold all through the Winter.

As regards what horses should be used to put with the mare, of course, that must be left to those who intend breeding. No matter what kind of mare a person may be going to breed from it is always best to obtain the best horse which can possibly be had. People who really understand this business well do not always go in for a certain horse because it has a good name, but they look out for one to match the mare as much as possible. I will endeavour to make myself more plain in this respect. Suppose a person has a breeding mare with a good body and legs, but low in the withers, he should look out for a horse very high in the withers. If it is a nag mare with a short, chubby neck, select a sire with a long neck. Always endeavour to obtain a horse which is fully developed in the points lacking in the mare. It is only by careful selection in this way that good animals are produced, and when a person takes care in choosing a horse the owner will often realise from £5 to £20 more for the colt when it is sold, or, at any rate, it will be that much more valuable.

I will mention one or two instances which have come under my notice. I have known nag mares rather short in bone and with a very heavy shoulder put with a sire heavy in the neck and shoulder also, which has resulted in the colts being far heavier than the mother, simply through using a horse which was fully developed in the same way as the mare. If a horse which has to work hard is too heavy in the shoulder and neck, no matter whether it is a hunter or a nag horse, it is almost sure to go over at knee. I have seen small cart mares fairly free from hair upon the legs trot like a pony, and when such animals as these are put with thorough-bred horses it is wonderful what colts they throw. Some of the best nags and hackneys are bred from a light cart mare and a thorough-bred horse. The only difficulty is they are sometimes a little wide in the jaw bone. Occasionally too, they are rather wide across the hips and hind quarters, showing a little too much of the mother, but this is not always the case. If they are bred from again with a good hackney they will throw some splendid stock.

I find many small farmers who have only two or three horses make a great mistake. If they want to breed from a mare, and they can get a horse lent them to put with her for about 30/- or £2 they will at once say that is the horse for their mare. But if they are asked £5 5s. for a good animal they say it is too much. This is a great mistake. In nine cases out of ten, when bred from thorough-bred horses, if the colts were sold at six years old they would make from £10 to £20 more than when an ordinary horse is used. At the same time it would not cost 1/- more to keep them. I am pleased to say farmers appear to be waking up to this fact more than they did a few years ago, and do not seem to mind paying a little more money as long as they can get a horse to suit their mare.

Now it sometimes happens that a gentleman gets hold of an extraordinarily good mare, which has plenty of pluck and durability, and the owner feels no money would purchase such an animal from him. In such a case as this I maintain it is foolishness on the part of the owner not to breed from such an animal. Even if he is living in a town the mare can be used up to the last six weeks or two months before she foals ; then if there is no accommodation at home, arrangements can easily be made with some farmer to foal the mare down and rear the colt. Most farmers are glad to rear a young colt for from £7 to £10 a year, according to the part of the country they may be living in. Some will do it for less than that. I have known colts to be reared in this way which have cost the owner from £35 to £45 up to when they have been from four and a half to five years of age, and the colts were worth from £70 to £100. Of course, I do not say they always turn out as well as this, there is always a certain amount of speculation in rearing young colts—especially nag colts. No one knows how they are going to turn out, they may show splints, side bones, or meet with an accident. A person must look on the bright side of things. If everyone were to say they did not like to breed, in case the colt should turn out badly, horses would soon become very scarce.

The question may arise here in the minds of some people—" At what age can a mare be bred from ? " Plenty of mares are put to the horse for the first time when they are from thirteen to fifteen years old, but at any age from three to fifteen will do. When I was a boy I remember a thorough-bred mare being put to a sire when

she was twenty-nine years of age, which had never had a
foal, and she threw a splendid colt in her 30th year. This
statement is perfectly true, for I knew the animal well and
rode on her back, and the colt's as well. The owner had
possessed the mare from when she was two years of age so
there was no mistake made in her age. The same year
a mare twenty-five years of age threw a colt on the same
farm, but it was not her first, it was the fifteenth she had
had.

It is perfectly safe to begin breeding from a mare
when she is from twelve to fifteen years of age—that is
to say, after the owner has had a good amount of work
out of her. I merely mention the fact of the old mares
breeding to show it is possible to get colts from them
even at the ages I have mentioned. I would strongly
recommend a gentleman who has got hold of a good
mare or two, noted for pluck and endurance, when he
has finished with them for carriage work, to breed from
them. If the owner would only do this he might get
some good horses. There is usually a difficulty in buying
them from horse dealers, as I suppose there is no class
of men who have such a bad reputation. I do
not mean to say there are no honest horse dealers,
because I know there are a few, but there are also many
dishonest ones. An inexperienced person can seldom
get taken in with anything more than in buying a horse,
as a man who understands his business well can get up a
horse so well for a time when it is for sale. This is one
reason why I decided to write a chapter on breeding.

If those who want to buy horses would be wise they would never get them from horse dealers unless they have them for a time on trial.   I know it is rather hard on those who deal in horses to advise this, but I have seen so many poor fellows, quite ignorant as to what a good animal should be, buy one for their own use, with money perhaps they have been saving up for years, and when they get the horse home it is fit for nothing but cats' meat.   I have come across many such cases as this, but will only just mention one which occurred as I am writing this chapter.   A poor man who had been saving up for years trusted a fellow who knew a good deal about horses to buy him one which would do him good service, and a very short time after he was ordered to have the poor thing killed.   The animal had never earned him a sovereign, and, though at the time the horse was bought it was nice enough to look at, yet the poor man did not have it more than four months, and it never worked more than about twice altogether. The money with which this man bought his horse took him eight years to save.   This is what I call hard lines, and just shows us how people can be taken in with horses.

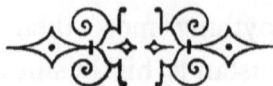

CHAPTER XI.

REARING YOUNG COLTS.

Treatment of Mares when suckling colts—The Colts and their treatment.

THERE are various opinions expressed with regard to rearing young colts   Some people will put a mare in constant work about a fortnight or three weeks after she has foaled, and of course she is nursing the foal at the same time.   Now it does not injure or keep back the colt much if the mare only works steadily.   For instance, if she is a cart mare it will do her no harm to let her do a few odd jobs on the farm.   In such a case as this I do not think it makes the slighest difference to the colt in any way.   Some farmers however, when they have a young colt, let the mare go to plough and work hard, this is both bad for the mare and very injurious to the colt ; as a rule the young ones do not get along nearly so well.

Then again, people who have never been accustomed to bring up a colt work the mare.   It may perhaps be a hackney or nag mare, and when she comes home she naturally wants to get to the foal, and the latter is in just the same mind.   Now the foal should never be allowed to suck when the mare is hot, not until she has cooled down.

Sometimes a good drop of her milk is wasted as it often runs to the ground. If it is really necessary to use her more after the foal is three or four months old I do not think it does much harm, but it is much better if the mare can be turned out to grass altogether, she naturally gives more milk than she does when she is working. It is just the same with young colts as it is with other young animals, the better they are nursed in the first instance the finer animals they turn out in the end, as there is a good foundation laid.

Foals usually get on better while they are with their mother than after they are weaned. I am often asked the question, "how long ought foals to be nursed by the mare ?" or in other words, "at what age should they be weaned ?" My answer is, "as long as the mare gives milk," which is usually from six to seven months, that is the time breeders usually allow the foals to suck. If they are allowed to remain with the mother six months they will take no hurt whatever. After the colt is taken from the mare many people turn it out in a grass field, or let it lie in a straw yard, and give it a little rough or mouldy hay to eat. In this case the colt makes a great deal of stomach, the hair grows very long on the body, and the animal becomes poor. How can anyone expect a young growing colt to get on well if it is fed on nothing but rough hay and a little grass all the Winter. If a person takes two colts of equal value from the mares in the Autumn, and feeds one in the way I have mentioned, and the other with a little corn, chaff, good hay, turnips

mangolds, carrots, or anything of that kind cut up and
mixed with its food, the latter animal will grow very fast all
through the Winter, and if both colts were sold the following
Spring, the one which had been fed well and looked after
properly would be worth from £5 to £7 more than the
other.

It is very bad policy to attempt to rear young colts on
poor inferior food during the winter months.   I have seen
it done in many instances, but always find the better food
they are fed upon through the Winter the more satisfaction
they give their owner in the end.   Of course young colts
pick up during the Summer and get quite fat when they
have plenty of grass, and sometimes they may not have
gone down much in condition if they are fed on rough food
the second year, but if the same experiment were tried again
the second Winter, even then, the colt which had been well
fed would be worth much more than the one which had been
brought up on rough diet.   The former will pay the owner
far more in comparison than the amount the extra food has
cost, especially if it is an entire colt.   This is not the only
benefit however which is derived from feeding the young
animals on good food, they grow so much better in shape,
and have more life and vigour in them when they are well
seen to.

Some young colts lie out the whole of the Winter and
have nothing but a little dry grass which they pick up in
the field, unless the snow covers the ground.   This is what
I call dragging out an existence.   I have no objection to
them lying out altogether the second year if it is a mild

Winter, but they should not do so the first year, they should be housed in a nice open shed. I do not believe in keeping young colts shut up in a warm place, because there is nothing like letting them have plenty of fresh air and keeping them in a nice even temperature. Young colts will do better if they are let run out, but they ought to have a little corn and good hay. Many breeders give them crushed oats and corn as soon as they can eat, right up to the time they are sold, or put to work, and the young horses pay for it afterwards, especially when they are well-bred animals. Some of my readers may say, " It does not pay to rear colts, as you can buy what you want," not always, it is sometimes rather difficult to do this. When you breed them yourself you can rear them to suit the purpose you require them for. But, says someone, " Occasionally they throw a splint, or are not perfect in some way, then it is a loss." Yes, that is quite right, there is always a little risk in breeding horses, just the same as there is in everything else. We must not expect every colt to be perfect. Even when you buy a horse which you think will suit you in every way, the animal does not always turn out equal to your expectations. There is a certain amount of risk whether you breed or buy a horse

During the winter months there is nothing like a little crushed linseed and oats for young colts. I do not say the quantity which should be given, because some people cannot afford to give much corn, another thing the colts do not require a great deal, it is a little they want. They are also very fond of bran, and will clear their chaff up nicely

if a little is given them.   Those who have valuable colts
should keep them in a large shed or box all through the
Winter and feed them well.   When a breeder is saving a
colt for an entire, before the mare has ceased to give milk,
he commences giving the young entire cows' milk to drink,
when it is warm, straight from the cow ; this of course
keeps the animal growing fast all through the Winter.

People often wonder how it is that the colts are got to
such a size for their age as many of them are at the shows.
Giving them cows' milk is the secret.   Of course a little
tonic, in the way of horse powder, put in their food
is a good thing, as it gives them a keen appetite and
they relish their food, though they do not always
require this treatment.   If they have the tonic too
often it does not have the desired effect.   My theory
is, if young horses will not pay for being well cared
for and fed properly, they will not pay for being reared
badly.   My advice is, keep them as well as your means will
allow.   I know there are poor farmers and others who
cannot afford to keep their colts well, but I say if you
can manage it, do so, as they will pay a good percentage in
the end on the money laid out on them.   As I am a
horse-breeder I speak from experience.

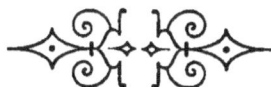

CHAPTER XII.

BREAKING IN.

How Colts are spoiled—conquer the horse or the horse will conquer you—Training to harness and in the Stable—Treatment of Jibbing Horses.

THERE are very few people who will go to the trouble of breaking in horses. Many of my readers, however, may during their lives have occasion to buy a horse which has not been broken in. Others again may go in for breeding a few young colts, so I thought a chapter on this subject would not be out of place. I have often said I would never breed horses, as I could buy them cheaper; but when one gets hold of an extra good mare they are tempted to breed from her, and this is how it was with me. This is really one of the most important periods in a horse's life. Thousands of splendid animals are spoiled through not being broken in properly. Many people may say '' Properly ! Yes, there comes the difficulty : what *is* being broken in properly? '' Different horse breakers have various methods of breaking in, and quite right too. One cannot go by certain rules in horse breaking, and put each colt under the same training. Horses vary so much in their temperaments, especially those which have

G

not been broken in. Some have very bad tempers when they are first handled, and others with a little coaxing can be made to do almost anything. When harsh measures are used the tempers of many colts are completely spoiled. On the other hand many horse breakers allow them to have their own way entirely, and when this is the case their tempers are never conquered.

I have known horses which have been broken in for two or three years, and to all appearance no one could do anything with them, they would only go where and when they liked, and stop when and where they liked. As I was brought up to horse breaking in my younger days I was able to judge the different dispositions of the animals, and have taken horses in hand, even these last few years, which the owners could do nothing with. These I have tamed down so that they would go splendidly. In other cases, I have seen horses that would not go, as well as those which required harsh treatment. No matter how bad tempered a horse is, it can usually be conquered if the person who has the management of it has the pluck to do it. I remember seeing a horse only lately which was bought in London at a very high figure. It was a very handsome animal and the gentleman's groom could do nothing with it. It had not been cleaned down for weeks, as both master and groom were too timid to do this. The horse would kick and bite as well, so they were both completely frightened of it. Another groom I knew went to see this horse, and when he went up to the animal the owner said he would be dashed to pieces.

He was a middle aged man, and when the horse came for him he hit him between the ears with a whip stock and knocked him down. When he was down he punished him severely. I believe it was the very next day the same man clipped him, and he stood as still as possible. After clipping him he attempted to ride him, but as soon as he got on his back he began to play his pranks. Then the man fetched him to the ground again and gave him another severe flogging, showing the horse plainly he was his master. The groom who conquered him bought him for a trifle for his master, and now both in riding and driving he can be done anything with. He is as quiet as a lamb, both in the stable and out. I mention this circumstance because it came so recently under my notice.

A horse is a very intelligent animal, and knows directly whether a person is frightened of it or not. Some require as much keeping in their place as an elephant. In breaking a horse it does not do to show the least fear. If the horse keeper once shows signs of nervousness the animal will very soon get the upper hand of him. I have seen many animals conquered in a similar way to the one I have referred to, and it is a most remarkable thing in almost every case where a horse shows such a fiery spirit, it turns out to be a splendid animal and a good trotter. It very often happens when such an animal as I have mentioned is broken in, it falls into the hands of people who soon become afraid of it. When this is the case the horse soon gets the upper hand of the attendant.

G2

It is quite right for a man to be kind to a horse, but at the same time he should be very firm. When a horse throws its ears back and lifts its foot up in the stable to kick, it should be spoken to very sharply and given a cut with the whip. Instead of doing this the attendant often runs out of the way, and the next time the horse lifts its foot up he will probably kick if not corrected. It is always better to use light measures than harsh, in breaking a horse in. The horse breaker should always carefully notice the disposition of the animal before using harsh means. For instance, the first time a young horse has its bridle on it is frightened, and some harsh trainers think it requires a lash round with the whip at once. This is a great mistake. When such strong measures are taken it often spoils the animal's temper. When the bridle is put on for the first time it is well to let an old horse go before the young one a few times, and the bridle of the latter should not have blinkers on, so that the animal may see what is going on. First the two should be walked across a grass field, then trotted, then run round a very big ring with the old horse in front. Just lay the whip across them if they are idle and will not run, but do not under any circumstances hit them hard at first, if so the younger horse becomes stubborn and will not move. Take them out next day and give them a good run again.

It is well when a young horse is being broken in, to stroke it and give it a little corn. After the bridle is put on the attendant should again speak kindly to the colt and give it a handful of corn at the same time. If this is done the animal soon begins to look upon you as its friend, instead of its enemy. This should be done every day while the colt is being broken in. The unfortunate part of the business is, people expect too much from young horses all at once. We call it breaking colts in, but I think it should be called educating them to do their duty. A young horse wants teaching one thing at a time. First, it should learn to walk nicely, and then to trot. After it can trot properly, a collar should be put on with hames in, so that the traces dangle down a little way and keep touching him, at the same time he should be spoken to kindly. The next thing to do is to tie some ropes on the traces, and some one should run behind, about six or seven yards away, to hold the ends. By doing this the traces touch the sides of the horse, and when it goes round the ring the traces should be pulled well on one side, so that they will rub against the horse. The next thing is to put a saddle on him, and a crupper to hang down. If the horse is quiet with that he can be cruppered, that is to say, the tail may be put through the crupper. The fore foot should be held up, so that he cannot kick, while a second man cruppers him. If the pupil runs well in harness, it can then be put in the cart, but of course great care should be taken. Before he is put in the cart he should have a bridle with blinkers on, otherwise when the wheels begin to go round it often startles the horse. When this is done, while the colt is standing quite still, the cart should be gently drawn up to him, and at the same time give him a little corn, and speak kindly to him, diverting his attention as much as possible.

When he is made secure do not attempt to get in the cart at once, only just lead him. I need hardly say that the tackle used for fastening him to the cart should be strong, in case he tries to get away. In this way he should be made to walk, or trot gently, round a grass field, because there is not so much noise from the wheels of the cart.

If the colt is intended principally for saddle work, after it has had a little exercise the saddle should be put on and the girth buckled very gently, not too tight, but just so that the saddle will not turn round. It is as well when a horse is broken in to break it in for harness as well as saddle work, then the animal is much more valuable if required for sale. Plenty of people like a good saddle horse, but cannot afford to keep two. If they are broken to harness when young they usually take to it years afterwards, when they are put to it, quietly. In all cases if a young horse is being taken out of the cart great care should be bestowed upon having every strap loose, so that there is nothing hanging to the cart, if so, the animal will most probably be spoiled. Carelessness on this point will get the colt in the way of rushing out of the trap before being properly released. In this case they are almost sure to break some part of the harness, and it will be a long time before they can be trusted again. When the colt will go along quietly by leading, the trainer should get into the cart, or trap, take the reins and teach the young horse how to be guided, by slightly pulling them. As a rule it takes from a month to six weeks to break a colt in properly, and the training should be carried out every day.

While these lessons are going on outside, the pupil should have a certain drill to go through inside the stable, until he has learned to go out properly, and move on one side when the attendant goes to feed and clean him out, at the same time he should be taught to stand with his fore feet well out from his hind ones.   This can be done by gently tapping the heels of the fore feet.   It is well to do this also while they are in harness, from the very first, as a horse looks so much better when he is taught to stand out properly.   If the owner is fond of his horse he can train it to do almost anything while young.   With pains and perseverance I have seen a horse trained so that it would put its fore feet out so far from his hind ones that its stomach almost touched the ground.   The owner could just put his leg over the horse's back without any trouble.   It would be a fine thing if horses were trained in this way for ladies and old gentlemen to mount.   When training a horse to put his feet out well, it is a good thing to walk them round a straw yard, with plenty of loose straw about, then each time the young horse steps he is obliged to lift his feet up high, which gives him a graceful appearance.

There would not be half the trouble in breaking young colts in, if the owner would go the trouble of putting a halter on them as soon as they were foaled, and leading them about till they were perfectly tame and docile.   All the colts I have had to do with

I have always brought up so that I could lead them anywhere in a hemp halter before they were broken in, and in not one single instance have I had any difficuly in training them. It is only a matter of teaching them to do their duty. I have gone to colts at two years old when they were out in the field, jumped on their back and galloped them about without any difficulty, and I have seen them put in a cart the first day they were handled. If young horses were brought up in this way from the very first, I do not think we should have many kickers, or horses which jib. Now, although I speak of kindness I wish to make myself plain in this respect. Occasionally it is necessary to be very strict with a young horse. When it shows a bad temper it does not want playing with, but correcting at once. This can only be done by those who have a good nerve ; a timid person should never attempt to break in a horse. In fact such people should have very little to do with them after they are broken in. If so they will very likely undo the good which has been done.

After a young nag horse has been broken in, there is often a difficulty in getting it to stand still, but such is not the case with young cart horses. As a rule they soon become calmed down as they are working with the old ones. Very often young horses which run in a trap will not stand while the owner gets up. They want training in this matter, as many of them which are sent to gentlemen on approval are returned because they will not stand long enough to allow anyone to get into the trap, or carriage. Then again, a spirited horse will scrape with its fore feet, and often rear, frightening the owner, who thinks it is going to throw him out. This kind of thing can easily be cured by putting a nose bag on the horse, with a little chaff and corn in it, and letting him stand for about ten minutes or a quarter of an hour, just long enough to eat half the feed of corn. If this is done from four to seven times, the horse gets to expect it, and usually there is no more trouble in this respect afterwards. It is a simple remedy, but I have never known it fail. I have trained horses in this way and have left them standing for two hours, covered up with a rug, of course. I do not say a young horse would do this, but an old one, which has been well trained, will stand for hours A little corn and a few kind words go a long way with horses, they appear to look for it.

I have not treated upon breaking horses in to run with a pair. This should never be attempted, unless the colt is fairly well broken in first, then he should be put side by side with a very steady old horse. I do not say one which will not trot, but one which can be trusted. When the colt has had a few runs with a well trained old horse, it can soon be trusted. I am often asked at what age a horse should be broken in. That depends on what it is required for. I have always found it best to break a young nag horse in at two or three years old, the latter age is usually the best. After they are well trained to harness, or saddle, they should be turned out another year and do no work, or, at any rate very little.   If they are allowed to run on a hard road they are almost sure to throw splints, or side bones.   Even after they are four years old they should be worked very steadily indeed.   In fact if they are really valuable horses they should be lightly worked for about a month, or six weeks, then turned out for another year, till they are five years old.   Even then they should not be allowed to do much running on the hard road till they are about six years old.

No doubt many of my readers will think this is a very ridiculous statement to make, but if a horse only does a little work before he reaches the age of six, he will be sounder on the legs when he is fifteen than many are at seven when they are worked at so early an age.   I know it seems a waste of time for a horse to do so little work by the time he is five years old, but if a person wants really good horses. the rules I have laid down shew the proper steps to take.   Seventeen out of every twenty horses, which have to run on the roads in large towns and cities, such as London, Liverpool, Manchester, Birmingham, and other large places, at four years old have quite done their work when they are from seven to ten years of age.   They are over at knee, and the sinews of their legs are drawn up like worn out old horses, through constantly being on the hard road before the bones of their legs are thoroughly set, and the sinews hardened.   Horses should never be used for that purpose until they are six years of age.   After they reach that age five years hard work on the stones will not hurt them half as much as two before they reach that age. Tramway companies, cab proprietors, and 'bus proprietors do not think of buying horses until they are six years of age, and their horses last as long again.

Now the question may be raised here by many of my readers "How can it pay to breed horses and keep them so many years without working them?" There is certainly a great deal in that, but a horse at six years old is worth more than a horse at four. A farmer can often utilize a horse a little on a farm, by putting a saddle on him and riding him round the farm, or taking him to market once a week. Such light work as this does not hurt the young horse, it does it good, but it should not be kept at hard regular work. As a rule, breeding cart horses pays much better than breeding nag horses, because cart colts can be broken in at two years old, and do a little light work if required. At three years of age they can go to constant work on the farm, as long as they are not put to draw any very heavy loads. Cart colts want but very little breaking in. The halter should be put round their neck and the colts led for a few times ; their harness can then be put on. If they are led round for a few times it will be sufficient. When there are three chain horses the young horse can be put in the middle and taken to plough. The colt will want leading by a man, and should not be trusted with a boy for the first month, as many farmers do, and after that, if a little care is bestowed upon them, they are but very little more trouble, and will work as quietly as old horses. Of course care should be taken when they are put in a cart. They should be allowed to draw it about once or twice without putting any load behind it. Where a great mistake is often made, is in putting a load in the cart too soon after the young horse is broken in, as when the colt comes to find the load does not move, which is often the case, the animal begins to jib, and can never be trusted again. When this is the case, at six years old they are not worth so much by from £10 to £30 as they would have been had they been treated properly.

Perhaps there is no other place in England where so many horses are noticed to jib, as at the foot of London Bridge, though it is not steep, there are seldom two hours in the day but what some horse turns his work up and has to be helped over the bridge by other horses. This jibbing is usually caused through proper care not being taken in breaking in. Carelessness on the part of the groom or driver sometimes will cause horses to jib, but it is usually through the breaking in; although injudicious treatment afterwards will sometimes spoil the temper, even if they are seven years old.

CHAPTER XIII.

DRIVING.

Some people's ideas about driving—The result of ignorance about driving—How fast to drive—Kindness and Common sense and the result of using them.

DRIVE a horse? Anyone can do that, one only has to hold the reins and steer straight. This is the opinion of many people. I suppose nothing will offend a person more than telling him he does not drive or manage his horse properly.

There is far more in driving than most people fancy, and many good horses are spoiled through careless driving. Some horses take a great deal more driving than others, inasmuch as their tempers vary so much. Some are very quiet in harness, good tempered, intelligent, and will go almost without any driving, but it is not so with all horses. Of course it is more difficult to drive a horse which runs than one which only walks, viz., a waggon or cart horse. My remarks in this chapter will therefore be principally confined to those which are driven in carriages and traps and have to trot. It may happen that a horse may be well broken in and sold for a good round figure, recommended as quiet in harness and a splendid goer. Very often on the agreement a horse will be entered as free from shying, quiet in harness, and sound in every respect, and before the new owner has had it a week it will shy and jump about the road at the sight of a bicycle or a piece of paper. Now this is very often the fault of the driver. Numbers of horses are completely spoiled simply for want of a little thought on the part of the person who is driving. I will endeavour to make it quite plain what I mean in this respect. A horse may be going comfortably along the road and suddenly sees something in his way which frightens him a little and causes him to cock his ears, or turn a little upon one side. The driver notices this and at once gives the animal a lash with the whip on the opposite side from which he shies. If you ask that person why he lashed the horse, he will probably tell you it was to prevent the animal shying. This is very wrong, as it adds fuel to the fire. Next time the animal sees anything he will not only be frightened of the object or sound but will be expecting a lash from the whip, which will probably cause him to do something desperate to escape this double source of uneasiness. Should it be a nervous horse and frisky as well, the consequences will probably be serious, as it will do almost anything to get out of the way if thoroughly frightened. If it be on a country road you may find yourself in a hedge or a ditch, if in a street on the pavement or through some shop window.

It is in this simple way that many valuable horses are spoiled after they are broken in. I am referring more particularly to carriage horses and hackneys. Never touch a horse with the whip when its shies. It must be borne in mind that a young horse is usually more or less nervous, and when it shies, or is a little frightened at anything, if the driver gives it a lash with the whip it naturally becomes very frightened. The person who drives says to himself, "Yes, I'll soon teach you to shy, my boy." But when anything on the road startles him again he is worse than he was the first time, because he is expecting every second to have the whip drawn across him. If this is done a few times the driver will have a great deal of trouble with it. If the horse belongs to a gentleman, his wife very soon tells him she dare not ride behind the animal, and he will have to get another horse. In this way many good horses are passed on from one to another. It cannot be wondered at that ladies are frightened to ride behind such horses, for the simple reason they are liable to turn the trap over. It is not always the animal's fault nor the person who sold it, it is merely through want of thought on the part of the driver. I am not saying this without experience, I have noticed horses treated in the way I have described over and over again. When a young horse shies he should never be punished under any consideration, but just held tightly by the reins and spoken kindly to. I always find it the safest plan to coax a horse which is nervous, and when he starts at anything pat him gently, saying "Whoa, boy, Whoa, my beauty," two or three times, and at the same time give him a handful of oats or bran. This should be repeated two or three times on the journey, especially when he shies at anything on the road. Many horses can be completely cured of this habit by speaking kindly. The animals get to look upon the person who treats them properly as their friend. I once bought a horse which would jump right round at the sight of a leaf or piece of paper blowing about, especially a bicycle. It was a very valuable horse were it not for that one fault, but it was sold on that account to run in a butcher's cart. I very soon cured him of the shying, certainly not in a week, nor a month, but in less than six weeks after I had him he was a different animal altogether in this respect. I found when he shied he would jump round, and then go as fast as he could, expecting a lash with the whip, but in less than three months I would have sat in the trap behind him without holding the reins. The horse would do anything I told him. He had been beaten till he was frightened at his own shadow, and I refused 100 guineas for him before he had been in my possession six months. When I had him he was between five and six, and he was valued then on account of his bad habits at about £15, but being cured of these tricks his value rose to 100 guineas in less than six months. I still have him at my place and have done for some years. This shews what kind and judicious treatment will do.

I have known many horses, which would jib and turn right round and back into a hedge or bank at the sight of almost anything on the road, sold at a nominal price, and when they have been treated properly, in a few weeks' time, they have been worth six times as much as they were sold for. This is simply owing to the new owner knowing how to drive and treat them properly. I am aware many young horses are spoiled in breaking in, I have dealt with this in the chapter on that subject. I am assuming now that the animals have been properly broken in. Many persons who are not accustomed to drive go along with too loose a rein and hold their hands too near their body, so that if the horse makes a stumble they have no power over it. A good horse driver, who understands his business well, keeps his hands out from the body, and if the horse shies or stumbles he has it well in hand. I do not mean to say a horse should always be held in very tightly, because some have a very tender mouth, while others have a very hard one In all cases they should be driven so that the driver can just feel the horse's mouth, and the reins should not be allowed to dangle on its back.

It is often a mystery to me why more people do not get their neck broken in driving, because some people let the horse have its own way entirely, and just jog along with the reins on the horse's back. The driver should not only hold the reins fairly tight, so as to have his horse well in hand, but should have one rein between the thumb and first finger, and the other between the large finger and third finger. Some people hold the reins between the thumb and first finger and the large and *second* finger, but those who do this should practise holding the reins in the way I have described, as it gives the driver more power over the horse.

H

I give an illustration, showing the position of the
hand and the reins between the fingers, so as to make this
plain.   I should also like to draw attention to another
mistake which is often made in driving, and which should
especially be guarded against.   Sometimes a horse will
stumble slightly through treading on a stone, or knocking
one leg against the other when trotting, at the same time it
may not fall.   When such is the case five people out of
every six will give the horse a sharp lash with the whip to
keep him upon his legs.   Now this is quite wrong.   When
a horse stumbles, if he is accustomed to being hit with the
whip, he gets very frightened in trying to save himself, and
will often go down, because he knows what he will get.
Instead of hitting the animal the driver should let him go
on twenty or thirty yards, then give him a touch up.   All
horses should be kept well in hand ; unless they are very
free indeed the driver should always let them know the
whip is behind them.   Some horses will not stand the
whip, these should have it gently drawn across them, it
keeps them up to their work.   Never on any account hit a
horse when it stumbles, it only makes matters worse.
Another thing too I notice where people go wrong in driving,
especially careless drivers, that is when the horse is going
very fast, sometimes they nearly run into another vehicle,
and instead of blaming themselves they beat the horse
unmercifully.   Now when this is the case, if the horse is at
all nervous, the next time he comes in close quarters he
will probably take the wheel off the passing trap or cart.

The above illustration shews the correct way of holding the reins and whip, so as to have good command over the horse. This position of the hands, with the body well forward, is the best for driving, and in case of emergency the safest.

It is a great pity that horses should have to be punished for what they cannot help, but so it is in many cases. Not only is it cruelty to the animal, but when a good driver comes to take him in hand he finds the poor thing is simply spoiled. Horses which are let out by job masters often have to suffer very much in this respect, especially when they are let out to a person to drive himself. Many people who hire vehicles are not accustomed to horses and the animals are driven unmercifully up hill and down, it is a matter of small consequence to the driver so long as the poor things keep going at a good pace. Anyone living near London can see a great deal of this, especially on Sundays. On one main road hired traps can be numbered from one hundred to three hundred if it is a fine Sunday, and taking into consideration the many roads there are, it just gives one some idea of the thousands of poor horses which are being driven to and fro in the way I have described every week. Greengrocers, grocers, costermongers, and all who are anxious to make money, will often let their horses out on a Sunday, but if they only knew how the poor things were used, I feel certain they would be more careful about letting them out. I do not say they are always treated badly, but when the party is going home, frequently the driver cannot see to drive straight, and the whole lot are lively with what they have had to drink, so the poor horse has to go as fast as it can, up hill and down, no matter whether he is tired out or not, and as for feeding it, many of them forget the animal eats anything at all. This is what I call cruelty to animals. One journey of this kind in a

week does more harm to the horse than six days' proper
work.   I hope this little book may fall into the hands
of some of those who let their horses out to hire for a
day's excursion, then perhaps it may induce some of
them to keep their animals at home, unless the owner
or someone who knows how to drive goes out with the
party himself.   It does not hurt a horse to do a long
journey if it is driven properly.   When a horse is going
a long distance, the driver should see it has a little
food every eight or twelve miles, and it should not be
driven as fast as if it were only going a short distance.
I have a weakness to drive fast myself if I am not
going far, but when I am going a long distance, I
never, under any consideration, drive more than eight
miles an hour.

When a person is on a long journey he should always
take some good corn, oats, a few split beans, peas, and
a little very good chaff, because when on the road there
are many little hotels and public houses which keep nothing
but a little bit of hay, and very often not even that.   But
if a person takes a little with him he can stop under a
green tree during the spring and summer months, and even
if only stopping for about five or ten minutes the nose bag
can be put on, and a rug thrown over the horse, which
helps the horse wonderfully.   Let him walk up the hills,
and trot steadily down, and he will take you along on the
level at a pretty good rate.   If a horse is hurried up the
hills when on a long journey he soon gets exhausted, unless
it happens to be a very hardy horse indeed, and even

then it is bound to tell on the animal. When a horse is driven properly and fed well he will do fifty miles in a day better than he would twenty-five if he were hurried along anyhow. It is a good plan always to carry a little oatmeal when on a long journey, to put in the water. This is a great stay to a horse. I mention this fact because many gentlemen now take their holidays by road with a horse and trap, because the seaside places are so full, and it is much nicer to go by road than by rail. There is something fresh to see every day, and, better still, change of air at the same time. I have taken my holidays for the last six years in this way, and can therefore speak from experience.

If a horse is driven properly and treated right it will do 500 miles easily in three weeks, this is only an average of about 20 miles a day, giving the horse a rest on Sundays. Sometimes, of course, the owner may have to drive forty or fifty miles in a day on account of there being no proper accommodation to put up at night. When a horse goes on a long journey like this it should always have its legs bandaged every night with flannel bandages. I prefer flannel bandages to linen, as they are not so likely to make the hair on the horse's legs ruck up. If the flannel is put on properly, and the legs are rubbed down nicely afterwards, it cannot be noticed. When linen bandages are used, unless they are kept very clean, they become hard and hurt the horse's legs. When the weather is hot, and the animal is on a long journey, it refreshes it very much if its legs are sponged down with

cold water, and the fore feet put in a pail of water
for a few minutes. This is not only refreshing to the
horse or horses, but it softens their hoofs a great deal.
When the roads are hot and dry their feet become very
brittle, but if they are put in water it makes the horn of
the hoof more elastic. I believe when the hoof of a
horse's foot is allowed to get so hard and brittle it is
the same as a person wearing a hard pair of boots, which
are not pliable to one's feet.

When anyone has a very stiff pair of boots on it
is difficult for that person to walk, but when they are
nice and easy—how different ; what a pleasure such an
one finds in running or walking. Now this is just the
case with the horse—the foot is inside the hoof, and
if the hoof is hard and brittle it often becomes contracted, and when this is the case it is very painful to
the horse. The hind feet do not suffer anything like
so much as the fore feet in this respect. It is well to
use a dressing for the hoof, which will be found in the
chapter on horses' feet, where I have dealt fully with
this subject. After a horse has been running it should
never be allowed to stand without having something
thrown across its loins, that is to say, it should not stand
for any length of time. A few minutes does not hurt
it. Some people make a great mistake in this respect
by throwing a cloth over a horse as soon as he stops,
after having been running fast for a good distance.
When the animal is all in a lather and a cloth is thrown
over him, it makes him perspire a great deal more.

The best way is to let him stand a few minutes till the steam goes off, then throw the rug on.   Of course if the weather is not very hot and the wind is cold, that is another thing ; throw the rug on as soon as the horse stops then, so as to prevent the animal taking a chill.

CHAPTER XIV.

WHAT IS THE BEST MATERIAL FOR THE STABLE FLOOR?

The difference in Litter—Paving: Bricks, Tiles, or Cement—Moss Peat Litter—Straw, &c.

THIS is a question often asked, but it is rather difficult to answer at all times. Much depends on what is going to be used upon the stable floor. If the latter is going to be kept covered with short material, especially moss peat, it makes but very little difference what the floor is made of. I often find farmers' stables, especially in Kent, paved with large flint stones. This is most absurd, because if a horse were restless and pulled all the bedding from under it, it would be liable to cuts its knees when lying down.

Bricks, too, are often used for the same purpose, especially in gentlemen's stables, so that they can easily be washed down, and at the same time they dry quickly. I do not like blue bricks myself for stables where there is nothing but wheat straw to litter the horses down with, as their feet slip too much. Some people use hard white bricks laid edgeways, which are very much better for the horses to get a good foothold.   They are of a much softer nature than the blue bricks.

During my travels the last few years I have found many stable floors, especially new ones, made of Portland cement.   This makes a fine hard bottom, and at the same time the horses can get a good hold with their feet when they want to get up.   Horses require a good firm, solid bottom under their feet, on account of their so often stamping, especially with their hind feet.   I would advise those who are building a stable to make a concrete floor, mixing in ground flint and cement, then put a facing of the best Portland cement and flint dust on that before it is dry.   If the floor is made in this way it will last a century as a rule without even cracking.   It not only has an advantage in being very durable, but there is nothing in it to hurt the horses' knees, and it gives them a good foothold as well.

In an ordinary stable, where a poor man only just keeps one horse and cannot spare much money to make a floor, if ordinary bricks are obtained and laid edgeways, and moss peat is put on the top of that, it will last a lifetime. Where bricks are used with nothing to cover them but straw, it is well to have a drain for the urine to pass away, with a large cup in it that fits easily, so that all the thick sediment can be taken out at least once a week.   If this is not seen to the drain soon becomes stopped up, as a good deal of the short manure gets washed down and sets very hard, unless there is a good flush of water used to keep the

the pipe clear. When the latter becomes stopped up, the drain soon begins to smell very badly, so much so that if you hold your head over the drain the ammonia is strong enough to cause you to shut your eyes.

I do not think it is necessary in this small work to dwell upon what farmers should have for their horses, as most of them have already got their stable floors in, and fresh farms are not cropping up as they used to a few years ago. One word, however, I may say to the farmers, that is, where they have so many rough stones for the stable they should lay either bricks down edgeways in the place of them in the way I have mentioned or concrete. The latter may perhaps come a little too expensive in some cases, but where one does not mind this it will last for centuries.

I am quite aware cart horses kick up the stable more than carriage horses. In the first place they are much heavier, and in the second place they itch a deal more in their heels and legs, especially where there is a greasy heel, or a rupture in the skin. They stamp with their feet very hard for ten or fifteen minutes together, and unless there is a very solid floor they knock it to pieces.

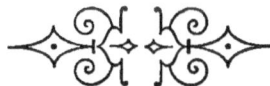

CHAPTER XV.

FEVER IN HORSES' FEET, AND WEAK JOINTS.

Fever in Horses' Feet—The Hunter: Shaken—The cold water cure—
Turn Horses out—The beneficial action of Dew—The Horse Dealer
"in Clover "—"Prevention is better than cure."

I HAVE made mention of fever in horses' feet in one or two places in this little work. This complaint is found frequently amongst horses, especially those which do a great deal of running on the hard roads. Hunters are very subject to it, sometimes jumping on hard ground brings it on, especially if the animal has corns. I do not say fever in the feet is always brought on through jarring a horse's foot or bruising the feet, but it is in many instances. A great many gentlemen will give a long figure for a good hunter, and after two or three seasons when the animal has been shaken about a great deal, of course it strains the horses' legs and joints : in some animals more than in others.

My intention is therefore to throw out a few hints so that when a horse does go wrong in this way the owner need not go to the expense of buying a fresh one, as many of them do. I know one farmer in Kent who makes a practice of taking in these used-up hunters year after year to grass. He has had some for as many as nine seasons. Sometimes when the animals are brought to the farm they can scarcely put one foot before the other, they are so used up, but when they go out in the Autumn their step is as elastic as that of a young horse.

For the benefit of those who may have horses in the condition I have mentioned, I will just state how this farmer treats them. Now it is a well known fact cold water is wonderfully strengthening for weak joints, and as soon as the horses are brought to the farm they are taken out and allowed to stand in a brook or stream for hours. The animal gets so accustomed to this kind of treatment, it looks forward to it with a great amount of pleasure and enjoys it very much. During the Summer very few of them want tying up in the water, they will stand in of their own accord for from three to six hours. There is no need to tie them up at all after they have been in about three times. If a horse that has weak joints, or is over at knees, as shown in illustration, is allowed to stand in water for from three to six hours a day, in almost every instance the legs become quite sound again and in many cases perfectly straight. Any kind of stream or pond will do to stand the animals in as long as the water comes over the horses' knees. When they are first put in the stream they ought not to remain there for more than an hour or two, especially in the early Spring, if so it is liable to bring on cramp in the legs, but after the animal gets accustomed to it, the time may gradually be increased to six hours out of the twenty-four. When a horse is undergoing this treatment it ought not to be worked, but turned out to grass. It is wonderful how Nature recovers itself when horses are turned out.

Some people say they do not like to turn horses out unless it is on damp ground. Naturally when the ground is damp it is better for the animal's feet, but it must be remembered the dew begins to come upon the grass even in the longest days of June, between seven and eight o'clock in the evening, and by nine o'clock the grass is quite wet. This can be tested by a person walking through the fields, he will soon find his boots wet through. Now dew is more penetrating than ordinary water, and when the horse is walking about all night, if the grass is fairly long, not only the feet but the lower joints of the legs also get wet. Very often the dew is not off the grass till nine in the morning unless the sun happens to be very warm indeed. The hotter the weather is the more dew we get. The grass is usually wet fourteen hours out of the twenty-four. It is not always wise to sell horses when they go wrong in the legs or have fever in the feet. If they are put under the treatment I have described, many of them become as sound in the legs as a young horse at four years old. Only within a few days of my penning these lines, I heard of a horse dealer who bought a horse for £19, and after he had put the animal in water in the way mentioned here for three months he sold him for one hundred guineas.

Of course those who live near the sea can stand the animals in the salt water, as that is a fine thing for horses, only it is rather difficult to tie them up, and another thing they cannot be left there for any length of time, because of the tide ebbing and flowing. I find ordinary water answers almost the same purpose, as it is more convenient. When they can stand in a pond or stream which is in a meadow, it is not only a question of the animal being benefited by standing in the water, but the dew from the grass is also very helpful to them, and the beautiful herbage which is found growing among the grass is really medicine for all cattle.

Before concluding this chapter there is one other thing I must not forget to mention. It often happens that those who have a good horse will work it hard, in some cases for years, without ever turning it out. Now this is a great mistake : if a good carriage horse or even a cab or 'bus horse were turned out as I have described in a good meadow and went under the water treatment for a month or six weeks every year, it would pay the owner fifty per cent., as it would last almost as long again, and what a delightful change for the poor horse. I have noticed carriage horses particularly, belonging to gentlemen, when the owner has been working them fairly hard and arranges so that they are turned out for a month or six weeks in the year, the animals always appear young and their legs keep perfect. This is only a reasonable request I am making for the horses. While it is a good thing for the animals it is a good investment for the owner. Horse dealers and men who are experienced in these things know perfectly well that many horses are bought from London and other large towns, turned out and put under similar treatment to that I have described here, and are sold for from six to ten times as much as they were bought for in the first instance. All I have said in this chapter may appear ridiculous to many people, but I have just given a simple and practical method of curing horses which go wrong in the joints or weak in the legs, and if gentlemen and owners of horses will act upon the

I

lines I have laid down, they will find they will have far less expense in buying fresh horses, but it will be conferring a great boon on the poor animals. Do not wait till the horse goes wrong in the legs before adopting the water plan—"prevention is better than cure."

I have given an illustration of a horse's front legs on a previous page, shewing how they appear after a long spell of work without relaxation, such as I am advocating. These legs would never have been in such a weak and strained condition had proper care been taken before they had gone so far.

FINIS.